P. S. Nair

Uncertainty in Multi-Source Databases

Springer

Berlin
Heidelberg
New York
Hong Kong
London
Milano
Paris
Tokyo

Studies in Fuzziness and Soft Computing, Volume 130

http://www.springer.de/cgi-bin/search_book.pl?series=2941

Editor-in-chief
Prof. Janusz Kacprzyk
Systems Research Institute
Polish Academy of Sciences
ul. Newelska 6
01-447 Warsaw
Poland
E-mail: kacprzyk@ibspan.waw.pl

Further volumes of this series can be found on our homepage

Vol. 112. Y. Jin
Advanced Fuzzy Systems Design and Applications, 2003
ISBN 3-7908-1537-3

Vol. 113. A. Abraham, L.C. Jain and J. Kacprzyk (Eds.)
Recent Advances in Intelligent Paradigms and Applications", 2003
ISBN 3-7908-1538-1

Vol. 114. M. Fitting and E. Orowska (Eds.)
Beyond Two: Theory and Applications of Multiple Valued Logic, 2003
ISBN 3-7908-1541-1

Vol. 115. J.J. Buckley
Fuzzy Probabilities, 2003
ISBN 3-7908-1542-X

Vol. 116. C. Zhou, D. Maravall and D. Ruan (Eds.)
Autonomous Robotic Systems, 2003
ISBN 3-7908-1546-2

Vol 117. O. Castillo, P. Melin
Soft Computing and Fractal Theory for Intelligent Manufacturing, 2003
ISBN 3-7908-1547-0

Vol. 118. M. Wygralak
Cardinalities of Fuzzy Sets, 2003
ISBN 3-540-00337-1

Vol. 119. Karmeshu (Ed.)
Entropy Measures, Maximum Entropy Principle and Emerging Applications, 2003
ISBN 3-540-00242-1

Vol. 120. H.M. Cartwright, L.M. Sztandera (Eds.)
Soft Computing Approaches in Chemistry, 2003
ISBN 3-540-00245-6

Vol. 121. J. Lee (Ed.)
Software Engineering with Computational Intelligence, 2003
ISBN 3-540-00472-6

Vol. 122. M. Nachtegael, D. Van der Weken, D. Van de Ville and E.E. Kerre (Eds.)
Fuzzy Filters for Image Processing, 2003
ISBN 3-540-00465-3

Vol. 123. V. Torra (Ed.)
Information Fusion in Data Mining, 2003
ISBN 3-540-00676-1

Vol. 124. X. Yu, J. Kacprzyk (Eds.)
Applied Decision Support with Soft Computing, 2003
ISBN 3-540-02491-3

Vol. 125. M. Inuiguchi, S. Hirano and S. Tsumoto (Eds.)
Rough Set Theory and Granular Computing, 2003
ISBN 3-540-00574-9

Vol. 126. J.-L. Verdegay (Ed.)
Fuzzy Sets Based Heuristics for Optimization, 2003
ISBN 3-540-00551-X

Vol 127. L. Reznik, V. Kreinovich (Eds.)
Soft Computing in Measurement and Information Acquisition, 2003
ISBN 3-540-00246-4

Vol 128. J. Casillas, O. Cordón, F. Herrera, L. Magdalena (Eds.)
Interpretability Issues in Fuzzy Modeling, 2003
ISBN 3-540-02932-X

Vol 129. J. Casillas, O. Cordón, F. Herrera, L. Magdalena (Eds.)
Accuracy Improvements in Linguistic Fuzzy Modeling, 2003
ISBN 3-540-02933-8

Premchand S. Nair

Uncertainty in Multi-Source Databases

Dr. Premchand S. Nair
Professor and Director of Computer Science
Creighton University
Omaha, NE 68178
USA
E-mail: psnair@creighton.edu

ISSN 1434-9922
ISBN 978-3-642-05705-2 e-ISBN 978-3-540-37099-4

Library of Congress Cataloging-in-Publication-Data applied for

A catalog record for this book is available from the Library of Congress.

Bibliographic information published by Die Deutsche Bibliothek
Die Deutsche Bibliothek lists this publication in the Deutsche Nationalbibliographie; detailed bibliographic data is available in the internet at <http://dnb.ddb.de>.

Springer-Verlag Berlin Heidelberg New York
a member of BertelsmannSpringer Science+Business Media GmbH
http://www.springer.de

Cover design: E. Kirchner, Springer-Verlag, Heidelberg
Printed on acid free paper 62/3020/M - 5 4 3 2 1 0

To my daughter Meera

PREFACE

Database and database systems have become an essential part of everyday life. In the course of a day, most of us perform several activities that involve some interaction with a database. All forms of banking activities such as deposit and withdrawal of funds, electronic transfer of funds, and automated bill payment involve accessing one or more databases. It is very easy to reserve airline tickets, hotel rooms, and rental cars from a cozy room at home or from any place that has a telephone or an internet connection. We trade stocks on line, pay our bills electronically, update our times sheets, make changes to various employee benefit programs using internet. The hidden power behind each of these conveniences is the technology that provides seamless access to one or more databases. Thus database is an integral part of our day to day life even though we may not be aware of it.

In the context of online shopping and product searching we can see product ranking and store ranking on many websites. Currently, products and services are ranked based on customer reviews. These reviews rate the quality of the product in a positive scale of 1 to 5. From an individual's perspective, internet provides a perfect vehicle for price comparison shopping in terms of convenience. From a business perspective, internet provides ample room for cost-cutting by automating more and more customer services. These trends call for the collection and storage of data from multiple sources. This in turn will place more demands on the capabilities of future database systems in collecting and integrating data from multiple sources with varying degrees of reliability. Therefore, database systems need to evolve in to decision making systems based on data from multiple sources having varying reliability. As a consequence, the next generation of database systems must deal with two major issues. The first issue is how to quantize the favorable (supporting) and unfavorable (opposing) facts so that they can be stored and processed efficiently. Secondly, how to use the reliability of the contributing sources in our decision makings. The objective of this work is to address these issues.

We introduce the concept of a confidence index set to mathematically model the above problem. A confidence index set (or ciset) is a generalization of the fuzzy set theory. In the case of fuzzy sets the membership is between 0 and 1. Thus one fuzzy set can be used to indicate the supporting (favorable) factor, say α. Similarly, a second fuzzy set can be used to represent the opposing (unfavorable) factor, say β. Thus if x is a fact, a tuple (x, β, α) can represent the complete information. Now, if x' is a logically opposing statement of x. Then what can we say about x'? Our intuition leads us to the fact (x', α, β). In the context of a database, we want to identify these two statements as one and the same. For this reason, we need to extend the fuzzy sets to ciset.

In Chapter 1, the concept of a ciset is introduced. The basic algebraic operations are presented. The proof that ciset is in fact a generalization of sets and fuzzy sets is also presented. A major contribution of this chapter to the fuzzy set theory is that, from the definition of difference operation in ciset theory, it is possible to define a meaningful difference operation in fuzzy set theory.

Chapter 2 is a simple introduction to relational database systems. This will allow anyone with no background in database theory fully appreciate the contents of this work. In Chapter 3, we present the ciset relational model. Chapter 4 is dedicated to extended relational operations in ciset relational database.

In Chapter 5, the semantics of ciset relational model is studied. Closely related notions of representation, possibility functions and alternate worlds have been the tools used by leading researchers to formalize the information content of databases with incomplete information. We use the notion of alternate worlds to formalize the information content of a ciset relational database. A ciset relation represents a set of (regular) relations known as alternate worlds. Once the alternate worlds are identified, a query on ciset relations can as well be processed against alternate worlds. Clearly, this approach is computationally inefficient and is not an approach we would recommend. On the other hand, this approach will well explain the semantics in a formal setting. In this chapter we prove that the ciset relational operations are precise.

Chapter 6 introduces additional ciset operations. These can be used in place of previously defined operations in certain applications. In Chapter 7, we summarize the information source tracking as suggested by Sadri. There are two main reasons for us to have a closer look at this approach. First, it is an important work in the sense that no one else has carried out a similar work that can keep track of the sources contributing the data. Thus, the approach is a major development in the area of multi-source database. Secondly, we can incorporate the ideas presented in this chapter to further extend the ciset relational database.

Premchand S. Nair

ACKNOWLEDGMENTS

The author is grateful to the editorial and production staffs of Springer-Verlag, especially Drs. Janusz Kacprzyk, Thomas Ditzinger, and Katharina Wetzel-Vandai. I am appreciative of the support of Dr. Timothy Austin, Dean, Creighton College of Arts and Sciences. I thank Dr. Albert Agresti, former Dean, Creighton College of Arts and Sciences. I am very grateful to my wife, Suseela, and my parents, Mr. Sukumaran Nair and Ms. Sarada Devi, for supporting my dreams.

CONTENTS

1
CISET

Fuzzy theory holds that many things in life are matters of degree. A black and white photo is not just black and white; there are many levels of gray shades that can be observed in a typical picture. As an example, a pixel can have a brightness value between 0 and 1. The value 0 may correspond to black and the value 1 may correspond to white. In this case, every number between 0 and 1 corresponds to a certain gray level. Fuzzy theory has made significant contributions to many areas. However, fuzzy theory is not sufficient in the area of information processing such as databases, expert systems, information retrieval and so on. The major drawback is its inability to capture negation of information. For example, let us say one of the CIA members has informed the headquarters that Facility X in country ABC is used to produce biological weapons. This information is assigned a fuzzy value 0.6. Later on, another source has informed the CIA headquarters that the facility is not used for the production of biological weapons. Currently there is no elegant way of keeping track of both pieces of information. One possible solution is to keep this new information with a fuzzy value assigned to it, say 0.3. However, above two facts are not treated as two pieces of information regarding the same subject. The concept of a ciset is introduced to integrate both supporting and opposing pieces of information. The term ciset refers to confidence index set and hence we start our discussion by introducing the confidence index.

Through out this book, L stands for a complete distributive lattice under a partial order $\leq$. Let $\vee$ and $\wedge$ be the join and meet operations on L. Further, we use 0 and 1 to denote the 0 and 1 of the lattice L. Whenever we present an example to illustrate various concepts, we use unit interval

$[0, 1]$ under the partial order less than or equal to ($\leq$) as L. Further, we shall use $\vee$ and $\wedge$ to represent the maximum and minimum operations on real numbers.

1.1 Confidence Index

Definition 1.1.1 *Let L be a complete distributive lattice and let $\alpha, \beta \in L$. Then a pair $a = \langle \alpha, \beta \rangle$ is called a* confidence index *on L.*

Example 1.1.2 *The following are examples of confidence indexes:*
$\langle 0,0 \rangle, \langle 1,0 \rangle, \langle 0,1 \rangle, \langle 1,1 \rangle, \langle 0.2,0 \rangle, \langle 1,0.4 \rangle, \langle 0.5,0.7 \rangle$.
Further, the following are not examples of confidence indexes:
$\langle 10,0 \rangle, \langle -1,0 \rangle, \langle 0,-1.2 \rangle, \langle 1.5,0.5 \rangle$.
Recall that in all our examples, L is assumed to be the unit interval.

Example 1.1.3 *The Facility X in country ABC is used to produce biological weapons has a confidence index $\langle 0.3, 0.6 \rangle$. To begin with, we have no information about the Facility X. As such the information "The Facility X in country ABC is used to produce biological weapons" has a confidence index $\langle 1, 0 \rangle$. Upon receiving supporting evidence from first source, evidence is evaluated and the β value is set as 0.6. In other words, the confidence index is modified to $\langle 1, 0.6 \rangle$. The second source now produces contradictory information. After carefully evaluating those pieces of information, the α value is set as 0.3. Thus the confidence index changes to $\langle 0.3, 0.6 \rangle$. Thus using the confidence index, both supporting and opposing pieces of information can be collected and presented in a unified manner.*

Definition 1.1.4 *Let $a = \langle \alpha, \beta \rangle$ be a confidence index. Then $l(a)$, the* lower index *of a, is α and $u(a)$, the* upper index *of a, is β.*

Two confidence indexes $a_i = \langle \alpha_i, \beta_i \rangle, i = 1, 2$ are equal if and only if $l(a_1) = l(a_2)$ and $u(a_1) = u(a_2)$. We now proceed to define $\prec, \preceq, \succ$ and $\succeq$. Confidence indexes $a_1 \prec a_2$, if $l(a_1) \geq l(a_2)$ and $u(a_1) < u(a_2)$ or $l(a_1) > l(a_2)$ and $u(a_1) \leq u(a_2)$. Of course $a_1 \preceq a_2$ if and only if either $a_1 \prec a_2$ or $a_1 = a_2$. Further, $a_1 \succ a_2$ if and only if $a_2 \prec a_1$ and $a_1 \succeq a_2$ if and only if $a_2 \preceq a_1$.

Example 1.1.5 *Let a be a confidence index. Then a can be plotted in the fourth quadrant of a two dimensional plane.*

FIGURE 1.1 The partial order $\preceq$.

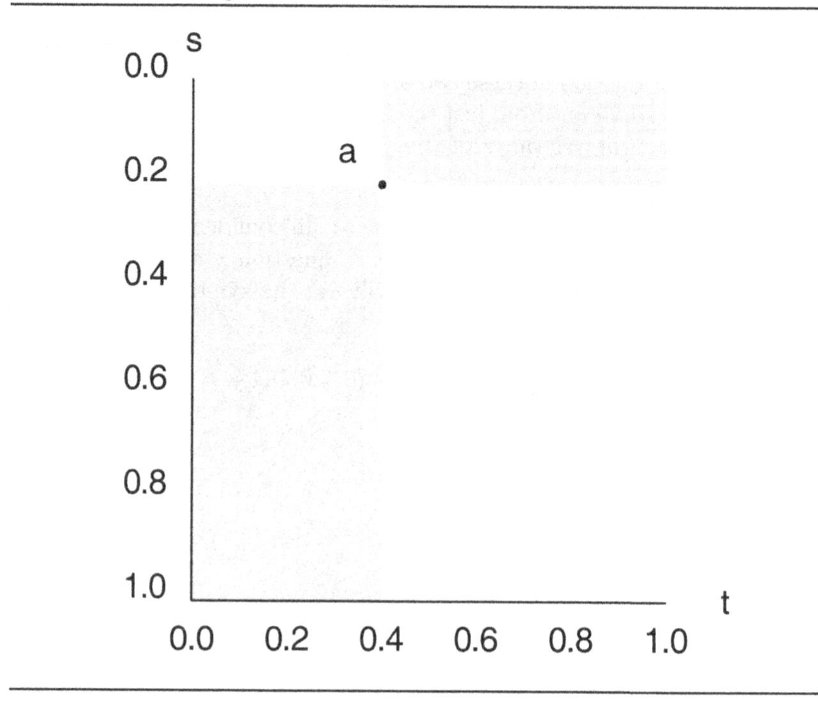

The shaded area to the right of a denote the set of all confidence indexes greater than or equal to a and the shaded area to the left of a denote the set of all confidence indexes less than or equal to a under the partial order $\preceq$.

We introduce four operations on the set of all confidence indexes. They are three binary operations *union* ($\cup$), *intersection* ($\cap$),*difference* ($-$); and one unary operation *negation* ($-$). Let $a_1 = \langle \alpha_1, \beta_1 \rangle, a_2 = \langle \alpha_2, \beta_2 \rangle$ be any two confidence indexes. Then

$$a_1 \cup a_2 = \langle \alpha_1 \wedge \alpha_2, \beta_1 \vee \beta_2 \rangle,$$

$$a_1 \cap a_2 = \langle \alpha_1 \vee \alpha_2, \beta_1 \wedge \beta_2 \rangle,$$

$$-a_1 = \langle \beta_1, \alpha_1 \rangle,$$

and

$$a_1 - a_2 = a_1 \cap (-a_2).$$

It is worth noticing that the set of confidence indexes is the Cartesian product of two lattices and as such is a lattice. However, with the introduction of negation and difference, there are more properties worth our investigation. Introduction of these two operations is what makes the set of confidence indexes different from just the Cartesian product of two lattices. The usefulness and the relevance of these operations will become evident in Chapter 4.

We use the notation $\mathfrak{C}(L)$ to denote the set of all confidence indexes on a complete lattice L. Further, if the lattice L in question is quite clear from the context or L can be any lattice, we will use the symbol $\mathfrak{C}$ instead of $\mathfrak{C}(L)$.

Example 1.1.6 *Let* $a_1 = \langle 0.3, 0.7\rangle, a_2 = \langle 0.5, 0.7\rangle, a_3 = \langle 0.4, 0.6\rangle, a_4 = \langle 0.2, 0.8\rangle \in \mathfrak{C}$. *Then we have the following:*
$a_4 \succ a_1 \succ a_2$ *and* $a_1 \succ a_3$,
$a_1 \cup a_4 = a_4$,
$a_1 \cap a_4 = a_1$,
$a_5 = a_2 \cup a_3 = \langle 0.4, 0.7\rangle \succ a_2$ *and* $a_5 \succ a_3$,
$a_6 = a_2 \cap a_3 = \langle 0.5, 0.6\rangle \prec a_2$ *and* $a_6 \prec a_3$,

FIGURE 1.2 Basic operations

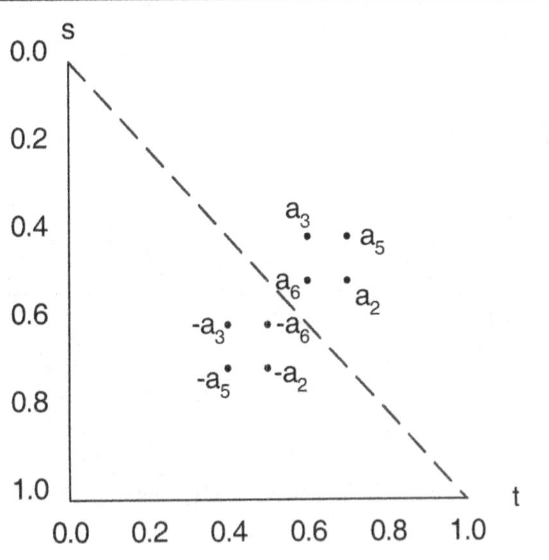

$\langle 0, 1\rangle - a_2 = \langle 0, 1\rangle \cap \langle 0.7, 0.5\rangle = \langle 0.7, 0.5\rangle = -a_2$,
$a_2 - \langle 1, 0\rangle = \langle 0.5, 0.7\rangle \cap \langle 0, 1\rangle = \langle 0.5, 0.7\rangle = a_2$,
$-(a_2 \cup a_3) = -\langle 0.4, 0.7\rangle$,

$= \langle 0.7, 0.4\rangle = \langle 0.7, 0.5\rangle \cap \langle 0.6, 0.4\rangle = (-a_2) \cap (-a_3)$,
$-(a_2 \cap a_3) = -\langle 0.5, 0.6\rangle$,
$= \langle 0.6, 0.5\rangle = \langle 0.7, 0.5\rangle \cup \langle 0.6, 0.4\rangle = (-a_2) \cup (-a_3)$.
It is worth noticing that $a_2 \cup a_3$ is at the top right corner of the rectangle defined by a_2 and a_3. Similarly, $a_2 \cap a_3$ is at the bottom left corner of the rectangle defined by a_2 and a_3. Further note that a confidence index such as a_2 and its negation $-a_2$ are symmetric with respect to the diagonal line joining $\langle 0,0\rangle$ and $\langle 1,1\rangle$.

Theorem 1.1.7 *Let $a_i \in \mathfrak{C}$ for $i = 1, 2, 3, 4$. Then we have the following:*

1. $a_1 \cup a_2 = a_2 \cup a_1$,
2. $a_1 \cap a_2 = a_2 \cap a_1$,
3. $a_1 \cup \langle 1,0\rangle = a_1$,
4. $a_1 \cap \langle 0,1\rangle = a_1$,
5. $a_1 \cup \langle 0,1\rangle = \langle 0,1\rangle$,
6. $a_1 \cap \langle 1,0\rangle = \langle 1,0\rangle$,
7. $a_1 \cup a_1 = a_1$,
8. $a_1 \cap a_1 = a_1$,
9. $a_1 \cup (a_2 \cup a_3) = (a_1 \cup a_2) \cup a_3$,
10. $a_1 \cap (a_2 \cap a_3) = (a_1 \cap a_2) \cap a_3$,
11. $a_1 \cup (a_2 \cap a_3) = (a_1 \cup a_2) \cap (a_1 \cup a_3)$,
12. $a_1 \cap (a_2 \cup a_3) = (a_1 \cap a_2) \cup (a_1 \cap a_3)$,
13. $-(a_1 \cup a_2) = (-a_1) \cap (-a_2)$,
14. $-(a_1 \cap a_2) = (-a_1) \cup (-a_2)$,
15. $(-(-a_1)) = a_1$,
16. *If* $a_1 \preceq a_2$ *then* $a_1 \cup a_2 = a_2$ *and* $a_1 \cap a_2 = a_1$,
17. *If* $a_1 \preceq a_3, a_2 \preceq a_4$ *then* $a_1 \cup a_2 \preceq a_3 \cup a_4$,
18. *If* $a_1 \preceq a_3, a_2 \preceq a_4$ *then* $a_1 \cap a_2 \preceq a_3 \cap a_4$.

Proof. Results are easy to prove and hence omitted. ■
It is worthwhile to note that the *law of contradiction* and the *law of excluded middle* do not hold. However, the reader may note that (13) and (14) above prove that *DeMorgan's Laws* hold.

Example 1.1.8 *Let* $a_1 = \langle 0.5, 0.7\rangle$ *and* $a_2 = \langle 0.6, 0.2\rangle$. *Then we have the following:*
$-a_1 = \langle 0.7, 0.5\rangle, -a_2 = \langle 0.2, 0.6\rangle$,
$a_1 - a_1 = \langle 0.7, 0.5\rangle = -a_1$,
$a_2 - a_2 = \langle 0.6, 0.2\rangle = a_2$,
$a_1 - a_2 = \langle 0.5, 0.7\rangle \cap \langle 0.2, 0.6\rangle = \langle 0.5, 0.6\rangle$,
$a_2 - a_1 = \langle 0.6, 0.2\rangle \cap \langle 0.7, 0.5\rangle = \langle 0.7, 0.2\rangle$.

Proposition 1.1.9 *Let* $a_i \in \mathfrak{C}$ *for* $i = 1, 2, 3$. *We have the following:*

1. $\langle 0, 1\rangle - a_1 = -a_1$,
2. $\langle 1, 0\rangle - a_1 = \langle 1, 0\rangle$,
3. $a_1 - \langle 1, 0\rangle = a_1$,
4. $a_1 - \langle 0, 1\rangle = \langle 1, 0\rangle$,
5. $(a_1 \cup a_2) - a_3 = (a_1 - a_3) \cup (a_2 - a_3)$,
6. $a_1 - (a_2 \cup a_3) = (a_1 - a_2) \cap (a_1 - a_3)$,
7. $(a_1 \cap a_2) - a_3 = (a_1 - a_3) \cap (a_2 - a_3)$,
8. $a_1 - (a_2 \cap a_3) = (a_1 - a_2) \cup (a_1 - a_3)$,
9. $a_2 - a_1 = (-a_1) - (-a_2)$.

Proof. Proof of first four statements are easy to follow and hence omitted.

5. $(a_1 \cup a_2) - a_3 = (a_1 \cup a_2) \cap (-a_3)$
$= (a_1 \cap (-a_3)) \cup (a_2 \cap (-a_3)) = (a_1 - a_3) \cup (a_2 - a_3)$.

6. $a_1 - (a_2 \cup a_3) = a_1 \cap (-(a_2 \cup a_3)) = a_1 \cap ((-a_2) \cap (-a_3))$
$= (a_1 \cap (-a_2)) \cap (a_1 \cap (-a_3)) = (a_1 - a_2) \cap (a_1 - a_3)$.

7. $(a_1 \cap a_2) - a_3 = (a_1 \cap a_2) \cap (-a_3)$
$= (a_1 \cap (-a_3)) \cap (a_2 \cap (-a_3)) = (a_1 - a_3) \cap (a_2 - a_3)$.

8. $a_1 - (a_2 \cap a_3) = a_1 \cap (-(a_2 \cap a_3)) = a_1 \cap ((-a_2) \cup (-a_3))$
$= (a_1 \cap (-a_2)) \cup (a_1 \cap (-a_3)) = (a_1 - a_2) \cup (a_1 - a_3)$.

9. First note that $(-a_1) - (-a_2) = (-a_1) \cap (-(-a_2)) = a_2 - a_1$.

■

1.2 Confidence Index Set

Let S be a set. A *confidence index set* or *ciset* (pronounced as see-set) is a mapping $F : S \to \mathfrak{C}$. One can think of F as assigning to each element $x \in S$, two degrees of confidence α and β such that α denotes the degree of confidence one has that $x \in S^c$, the complement of S; and β denote the degree of confidence one has that $x \in S$.

Let $S = \{u, v, x, y, z\}$. Define $F : S \to \mathfrak{C}$ as follows: $F(u) = \langle .4, .8\rangle, F(v) = \langle 0, 1\rangle, F(x) = \langle 1, 0\rangle, F(y) = \langle 0, 0\rangle, F(z) = \langle 1, 1\rangle$. Note that it is possible for an element be in S^c and S with same level of confidence and the sum of the levels of confidence of an element w defined as $l(F(w)) + u(F(w))$ need not be 1; rather it can be any value between 0 (as in y) and 2 (as in z). At first, it may strike as a contradiction. Since ciset is not asserting the membership of an element in S^c or S, but rather asserting the confidence levels obtained through various sources, there is no contradiction. For example, S may be a set of faculty members and F may be the performance of the faculty, as evaluated by their students. It is quite possible that some students may find a certain faculty excellent while other students may find the same faculty the worst teacher they ever had. Similarly, Internet auction sites maintain a rating system for its patrons. The same person is rated both positively as well as negatively by different patrons.

We say two cisets F and G on a set S are equal, and write $F = G$, if $F(x) = G(x)$ for all $x \in S$.

Definition 1.2.1 *Let F and G be two cisets on a set S such that $F(x) \preceq G(x)$ for all $x \in S$, then F is said to be* subset *of G and G is said to be a* superset *of F. If F is a subset of G and there exists at least one $x \in S$ such that $F(x) \prec G(x)$ then F is said to be* proper subset *of G and G is said to be a* proper superset *of F.*

If F is a subset of G, then F is said to be contained in G. Symbolically, we write $F \subseteq G$. If F is a proper subset of G, then we say F is strictly contained in G and is denoted by $F \subset G$.

Proposition 1.2.2 *Let F, G, H be cisets on a set S. Then*

1. $F \subseteq F$,

2. If $F \subseteq G$ and $G \subseteq H$, then $F \subseteq H$,

3. If $F \subseteq G$ and $G \subset H$, then $F \subset H$,

4. If $F \subseteq G$ and $F \nsubseteq H$, then $G \nsubseteq H$, where $\nsubseteq$ means "is not contained in,"

5. $F = G$ if and only if $F \subseteq G$ and $G \subseteq F$.

Proof.

1. For all $x \in S, F(x) \preceq F(x)$. Thus $F \subseteq F$.

2. For all $x \in S, F(x) \preceq G(x) \preceq H(x)$. Thus $F \subseteq H$.

3. For all $x \in S, F(x) \preceq G(x) \prec H(x)$. Thus $F \subset H$.

4. For all $x \in S, F(x) \preceq G(x)$ and there exists an element $y \in S$ such that $F(y) \succ H(y)$. Thus $H(y) \prec F(y) \preceq G(x)$. Thus $G \not\subseteq H$.

5. $F = G \Leftrightarrow$ for all $x \in S, F(x) = G(x) \Leftrightarrow$ for all $x \in S, F(x) \preceq G(x)$ and $F(x) \succeq G(x) \Leftrightarrow F \subseteq G$ and $G \subseteq F$. ∎

A ciset F on S is said to be an *empty ciset* if $F(x) = \langle 1, 0 \rangle$ for all $x \in S$. We shall use $F_\emptyset$ to denote the empty set. A ciset F on S is said to be the *universe ciset* if $F(x) = \langle 0, 1 \rangle$ for all $x \in S$. We use $F_{U(S)}$ to denote the universe ciset. Further, if there is no confusion, we shall abbreviate $F_{U(S)}$ to F_U. A ciset F on S is said to be a *singleton* if there exists a $y \in S$ such that $F(x) = \langle 1, 0 \rangle$ for all $x \in S - \{y\}$ and $F(y) = \langle 0, 1 \rangle$. We use I_y to denote the singleton. That is, $I_y(x) = \langle 1, 0 \rangle$ for all $x \in S - \{y\}$ and $I_y(y) = \langle 0, 1 \rangle$.

1.3 Basic Operations

In this section we introduce five operations on cisets. They are union, intersection, difference, product and complement. These operations allow us to construct new cisets from given cisets. We shall also study relationships among these operations.

Definition 1.3.1 *Let S be a set and let F, G be two cisets on S. The* union *of F and G, denoted by $F \cup G$ is a mapping from $S \to \mathfrak{C}$, defined by $(F \cup G)(x) = F(x) \cup G(x)$, for all $x \in S$.*

Clearly $F \cup G$ is the smallest ciset containing both F and G. In other words, if T is a ciset such that T contains both F and G, then T contains $F \cup G$.

Definition 1.3.2 *Let S be a set and let F, G be two cisets on S. The* intersection *of F and G, denoted by $F \cap G$ is a mapping from $S \to \mathfrak{C}$, defined by $(F \cap G)(x) = F(x) \cap G(x)$, for all $x \in S$.*

It may be noted that $F \cap G$ is the largest subset of both F and G. In other words, if T is a ciset such that T is a subset of both F and G, then $T \subseteq F \cap G$. Two cisets F and G on S are said to be *disjoint* if $F \cap G = F_\emptyset$, the empty set.

Definition 1.3.3 *Let S be a set and let F, G be two cisets on S. The* difference *of F and G, denoted by $F - G$ is a mapping from $S \to \mathfrak{C}$, defined by $(F - G)(x) = F(x) - G(x)$, for all $x \in S$.*

Definition 1.3.4 *Let S be a set and let F, G be two cisets on S. The* Cartesian product *of F and G, denoted by $F \times G$ is a mapping from $S \times S \to \mathfrak{C}$, defined by $(F \times G)(x, y) = F(x) \cap G(y)$, for all $(x, y) \in S \times S$.*

Definition 1.3.5 *Let S be a set and let F be a ciset on S. The* complement *of F, denoted by $-F$ is a mapping from $S \to \mathfrak{C}$, defined by $(-F)(x) = -F(x)$.*

Theorem 1.3.6 *Let F, G, H, J and K be any five cisets on a set S. Then we have the following:*

1. $F \cup F = F$,
2. $F \cap F = F$,
3. $F \cup G = G \cup F$,
4. $F \cap G = G \cap F$,
5. $F \cup F_\emptyset = F$,
6. $F \cap F_U = F$,
7. $F \cup F_U = F_U$,
8. $F \cap F_\emptyset = F_\emptyset$,
9. $F \cup F = F$,
10. $F \cap F = F$,
11. $F \cup (G \cup H) = (F \cup G) \cup H$,
12. $F \cap (G \cap H) = (F \cap G) \cap H$,
13. $F \cup (G \cap H) = (F \cup G) \cap (F \cup H)$,
14. $F \cap (G \cup F) = (F \cap G) \cup (F \cap H)$,
15. $-(F \cup G) = (-F) \cap (-G)$,
16. $-(F \cap G) = (-F) \cup (-G)$,
17. $(-(-F)) = F$,
18. *If* $F \leq G$ *then* $F \cup G = G$ *and* $F \cap G = F$,
19. *If* $F \subseteq K, G \subseteq J$ *then* $F \cup G \subseteq K \cup J$,

20. If $F \subseteq K, G \subseteq J$ *then* $F \cap G \subseteq K \cap J$.

Proof. Proof of each result follows from the corresponding result on confidence index. For example to prove (15), assume that $x \in S$. Then $(-(F \cup G))(x) = -((F \cup G)(x)) = -(F(x) \cup G(x)) = (-F(x)) \cap (-G(x)) = (-F)(x) \cap (-G)(x) = (-F) \cap (-G)(x)$. ■

It is worthwhile to note that the properties $F \cap (-F) = F_\emptyset$ and $F \cup (-F) = F_U$ do not hold in general. In logic, the former property is known as the *law of contradiction* and the latter property is called the *law of excluded middle.*

The reader may also note that (15) and (16) above prove *DeMorgan's Laws* do hold.

Proposition 1.3.7 *Let* $F, G,$ *and* H *be any three cisets on a set* S. *Then we have the following:*

1. $(F \cup G) - H = (F - H) \cup (G - H)$,
2. $F - (G \cap H) = (F - G) \cup (F - H)$,
3. $(F \cap G) - H = (F - H) \cap (G - H)$,
4. $F - (G \cup H) = (F - G) \cap (F - H)$,
5. $G - F = (-F) - (-G)$.

Proof. Proof of each result follows from the corresponding result on confidence index. ■

A ciset can be considered a generalization of set. Let S be a set and A be any subset of S. Define a mapping $F_A : S \to \mathfrak{C}$ by $F_A(x) = \langle 0, 1\rangle$ if $x \in A$ and $F_A(x) = \langle 1, 0\rangle$ if $x \notin A$. Thus a ciset can be considered a generalization of set. Similarly, if $\mu : S \to \langle 0, 1\rangle$ is a fuzzy subset of S, define a mapping $F_\mu : S \to \mathfrak{C}$ by $F_\mu(x) = \langle 1 - \mu(x), \mu(x)\rangle$, for all $x \in S$. This convention and notation is followed for the rest of this book.

Theorem 1.3.8 *Let* S *be a set and* $A, B \subseteq S$. *Then*

1. $F_A \cup F_B = F_{A \cup B}$,
2. $F_A \cap F_B = F_{A \cap B}$,
3. $F_A - F_B = F_{A - B}$,
4. $F_A \times F_B = F_{A \times B}$,
5. $(-F_A) = F_{A^c}$.

Proof. Let $x \in S$. The proof can be summarized as follows. Proof of first three results follows from the following three tables. We distinguish four cases: (1) $x \in A, x \notin B$, (2) $x \notin A, x \in B$, (3) $x \in A, x \in B$ and (4) $x \notin A, x \notin B$.

As an illustration, we prove result (1) case (1) as follows: Assume that $x \in A, x \notin B$. Then $F_A(x) = \langle 0,1\rangle, F_B(x) = \langle 1,0\rangle$. Thus $(F_A \cup F_B)(x) = F_A(x) \cup F_B(x) = \langle 0\wedge 1, 1\vee 0\rangle = \langle 0,1\rangle$. Now $x \in A\cup B$. Therefore, $F_{A\cup B}(x) = \langle 0,1\rangle$. Therefore, $F_A \cup F_B = F_{A\cup B}$.

	F_A	F_B	$F_A \cup F_B$	$F_{A\cup B}$
$x \in A, x \notin B$	$\langle 0,1\rangle$	$\langle 1,0\rangle$	$\langle 0,1\rangle$	$\langle 0,1\rangle$
$x \notin A, x \in B$	$\langle 1,0\rangle$	$\langle 0,1\rangle$	$\langle 0,1\rangle$	$\langle 0,1\rangle$
$x \in A, x \in B$	$\langle 0,1\rangle$	$\langle 0,1\rangle$	$\langle 0,1\rangle$	$\langle 0,1\rangle$
$x \notin A, x \notin B$	$\langle 1,0\rangle$	$\langle 1,0\rangle$	$\langle 1,0\rangle$	$\langle 1,0\rangle$

	F_A	F_B	$F_A \cap F_B$	$F_{A\cap B}$
$x \in A, x \notin B$	$\langle 0,1\rangle$	$\langle 1,0\rangle$	$\langle 1,0\rangle$	$\langle 1,0\rangle$
$x \notin A, x \in B$	$\langle 1,0\rangle$	$\langle 0,1\rangle$	$\langle 1,0\rangle$	$\langle 1,0\rangle$
$x \in A, x \in B$	$\langle 0,1\rangle$	$\langle 0,1\rangle$	$\langle 0,1\rangle$	$\langle 0,1\rangle$
$x \notin A, x \notin B$	$\langle 1,0\rangle$	$\langle 1,0\rangle$	$\langle 1,0\rangle$	$\langle 1,0\rangle$

	F_A	F_B	$F_A - F_B$	F_{A-B}
$x \in A, x \notin B$	$\langle 0,1\rangle$	$\langle 1,0\rangle$	$\langle 0,1\rangle$	$\langle 0,1\rangle$
$x \notin A, x \in B$	$\langle 1,0\rangle$	$\langle 0,1\rangle$	$\langle 1,0\rangle$	$\langle 1,0\rangle$
$x \in A, x \in B$	$\langle 0,1\rangle$	$\langle 0,1\rangle$	$\langle 1,0\rangle$	$\langle 1,0\rangle$
$x \notin A, x \notin B$	$\langle 1,0\rangle$	$\langle 1,0\rangle$	$\langle 1,0\rangle$	$\langle 1,0\rangle$

Proof of result (4) follows from the following table. We distinguish four cases: (1) $x \in A, y \notin B$, (2) $x \notin A, y \in B$, (3) $x \in A, y \in B$ and (4) $x \notin A, y \notin B$.

	F_A	F_B	$F_A \times F_B$	$F_{A\times B}$
$x \in A, y \notin B$	$\langle 0,1\rangle$	$\langle 1,0\rangle$	$\langle 1,0\rangle$	$\langle 1,0\rangle$
$x \notin A, y \in B$	$\langle 1,0\rangle$	$\langle 0,1\rangle$	$\langle 1,0\rangle$	$\langle 1,0\rangle$
$x \in A, y \in B$	$\langle 0,1\rangle$	$\langle 0,1\rangle$	$\langle 0,1\rangle$	$\langle 0,1\rangle$
$x \notin A, y \notin B$	$\langle 1,0\rangle$	$\langle 1,0\rangle$	$\langle 1,0\rangle$	$\langle 1,0\rangle$

To prove the result (5), we distinguish two cases: (1) $x \in A$ and (2) $x \notin A$. The proof follows from following table.

	F_A	$(-F_A)$	F_{A^c}
$x \in A$	$\langle 0,1\rangle$	$\langle 1,0\rangle$	$\langle 1,0\rangle$
$x \notin A$	$\langle 1,0\rangle$	$\langle 0,1\rangle$	$\langle 0,1\rangle$

■

Theorem 1.3.9 *Let S be a set and let μ, σ be two fuzzy subsets on S.*

1. $F_\mu \cup F_\sigma = F_{\mu\cup\sigma}$,

2. $F_\mu \cap F_\sigma = F_{\mu\cap\sigma}$,

3. $F_\mu \times F_\sigma = F_{\mu\times\sigma}$,

4. $(-F_\mu) = F_{\mu^c}$.

Proof. Let $x \in S$. Proof of first two results follows from the following two tables. We distinguish between two cases: (1) $\mu(x) \leq \sigma(x)$ and (2) $\mu(x) > \sigma(x)$. The first row of each of the following first two tables will correspond to the first case and the second row of each of the following first two tables will correspond to the second case respectively.

As an illustration, we prove result (1) case (1) as follows: Assume that $\mu(x) \leq \sigma(x)$. Then $F_\mu(x) = \langle 1-\mu(x), \mu(x)\rangle, F_\sigma(x) = \langle 1-\sigma(x), \sigma(x)\rangle$. Thus $(F_\mu \cup F_\sigma)(x) = F_\mu(x) \cup F_\sigma(x) = \langle (1-\mu(x)) \wedge (1-\sigma(x)), \mu(x) \vee \sigma(x)\rangle = \langle 1-\sigma(x), \sigma(x)\rangle$. Now $(\mu \cup \sigma)(x) = \mu(x) \cup \sigma(x) = \sigma(x)$. Hence $F_{\mu\cup\sigma}(x) = \langle 1-\sigma(x), \sigma(x)\rangle$. Therefore, $F_\mu \cup F_\sigma = F_{\mu\cup\sigma}$.

F_μ	F_σ	$F_\mu \cup F_\sigma$
$\langle 1-\mu(x), \mu(x)\rangle$	$\langle 1-\sigma(x), \sigma(x)\rangle$	$\langle 1-\sigma(x), \sigma(x)\rangle$
$\langle 1-\mu(x), \mu(x)\rangle$	$\langle 1-\sigma(x), \sigma(x)\rangle$	$\langle 1-\mu(x), \mu(x)\rangle$

$F_{\mu\cup\sigma}$
$\langle 1-\sigma(x), \sigma(x)\rangle$
$\langle 1-\mu(x), \mu(x)\rangle$

F_μ	F_σ	$F_\mu \cap F_\sigma$
$\langle 1-\mu(x), \mu(x)\rangle$	$\langle 1-\sigma(x), \sigma(x)\rangle$	$\langle 1-\mu(x), \mu(x)\rangle$
$\langle 1-\mu(x), \mu(x)\rangle$	$\langle 1-\sigma(x), \sigma(x)\rangle$	$\langle 1-\sigma(x), \sigma(x)\rangle$

$F_{\mu\cap\sigma}$
$\langle 1-\mu(x), \mu(x)\rangle$
$\langle 1-\sigma(x), \sigma(x)\rangle$

Proof of result (3) follows from the following table. We distinguish between two cases: (1) $\mu(x) \leq \sigma(y)$ and (2) $\mu(x) > \sigma(y)$. The first row of the following table will correspond to the first case and the second row of the following table will correspond to the second case respectively.

F_μ	F_σ	$F_\mu \times F_\sigma$
$\langle 1-\mu(x), \mu(x)\rangle$	$\langle 1-\sigma(y), \sigma(y)\rangle$	$\langle 1-\sigma(y), \mu(x)\rangle$
$\langle 1-\mu(x), \mu(x)\rangle$	$\langle 1-\sigma(y), \sigma(y)\rangle$	$\langle 1-\mu(x), \sigma(y)\rangle$

$F_{\mu\times\sigma}$
$\langle 1-\mu(x), \mu(x)\rangle$
$\langle 1-\sigma(y), \sigma(y)\rangle$

The proof of (4) follows from following table.

F_μ	$(-F_\mu)$	F_{μ^c}
$\langle 1-\mu(x), \mu(x)\rangle$	$\langle \mu(x), 1-\mu(x)\rangle$	$\langle \mu(x), 1-\mu(x)\rangle$

■

Given a ciset $F : S \to \mathfrak{C}$, it is possible to produce various subsets and fuzzy subsets of S. In particular, we introduce the following. Let $a = (s,t)$

be an ordered pair of real numbers. Define *a–cut set*, F^t_s, by $F^t_s = \{x \in S \mid u(F(x)) \geq t \text{ and } l(F(x)) < s\}$.

This definition is motivated by the fact that quite often, one may be interested in obtaining information having a certain level of confidence in the positive factor while like to limit the negative factor to a certain limit. In fact, this observation was behind our definition of the partial order $\preceq$ as well as the basic operations union, intersection, complement and difference. Other approaches are possible and are explored in Chapter 6.

Proposition 1.3.10 *Let $a_1 = (s_1, t_1)$ and $a_2 = (s_2, t_2)$ be two ordered pairs of real numbers such that $s_1 \geq s_2$ and $t_1 \leq t_2$ and let F be a ciset. Then $F^{t_1}_{s_1} \supseteq F^{t_2}_{s_2}$. In particular, if a_1 and a_2 are two confidence indexes such that $a_1 \preceq a_2$ then $F^{t_1}_{s_1} \supseteq F^{t_2}_{s_2}$.*

Proof. Let $x \in S$. Assume $x \in F^{t_2}_{s_2}$. Then $u(F(x)) \geq t_2$ and $l(F(x)) < s_2$. Therefore, $u(F(x)) \geq t_1$ and $l(F(x)) < s_1$. Hence $x \in F^{t_1}_{s_1}$. ■

Proposition 1.3.11 *Let $a = (s, t)$ be an ordered pair of real numbers and let F, G be two cisets. Then $(F \cup G)^t_s \supseteq F^t_s \cup G^t_s$ and $(F \cap G)^t_s = F^t_s \cap G^t_s$.*

Proof. Clearly, $(F \cup G)^t_s \supseteq F^t_s$ and $(F \cup G)^t_s \supseteq G^t_s$. Therefore, $(F \cup G)^t_s \supseteq F^t_s \cup G^t_s$. Let $x \in S$. Now, $x \in (F \cap G)^t_s \Leftrightarrow u((F \cap G)(x)) \geq t$ and $l((F \cap G)(x)) < s \Leftrightarrow u(F(x)) \geq t$ and $l(F(x)) \geq s$; and $u(G(x)) < t$ and $l(G(x)) < s \Leftrightarrow x \in F^t_s \cap G^t_s$. ■

We use the symbol F^t to denote the set $\{x \in S \mid u(F(x)) \geq t\}$. Similarly, we use the symbol F_s to denote the set $\{x \in S \mid l(F(x)) < s\}$. The *upper fuzzy set* corresponding to the ciset F is defined as $\overline{F} = \{(x, u(F(x))) \mid x \in S\}$. In other words, $\overline{F}$ is a mapping from S into $[0, 1]$ such that $\overline{F}(x) = u(F(x))$. Similarly, the *lower fuzzy set* corresponding to the ciset F is defined as $\underline{F} = \{(x, l(F(x))) \mid x \in S\}$. In other words, $\underline{F}$ is a mapping from S into $[0, 1]$ such that $\underline{F}(x) = l(F(x))$.

Theorem 1.3.12 *Let S be a set and $A, B \subseteq S$. Then*

1. $(F_A \cup F_B)^1 = (F_A)^1 \cup (F_B)^1 = (F_{A \cup B})^1$
2. $(F_A \cap F_B)^1 = (F_A)^1 \cap (F_B)^1 = (F_{A \cap B})^1$
3. $(F_A - F_B)^1 = (F_A)^1 - (F_B)^1 = (F_{A-B})^1$
4. $(F_A \times F_B)^1 = (F_A)^1 \times (F_B)^1 = (F_{A \times B})^1$
5. $(-F_A)^1 = (F_{A^c})^1$

Proof. Let $x \in S$. The proof can be summarized as follows. Proof of first three results follows from the following three tables. We distinguish four cases: (1) $x \in A, x \notin B$, (2) $x \notin A, x \in B$, (3) $x \in A, x \in B$ and (4) $x \notin A, x \notin B$.

As an illustration, we prove result (1) case (1) as follows: Assume that $x \in A, x \notin B$. Then $F_A(x) = \langle 0,1\rangle, F_B(x) = \langle 1,0\rangle$. Consequently, $x \in (F_A)^1$, $x \notin (F_B)^1$ and thus $x \in (F_A)^1 \cup (F_B)^1$. Also, $(F_A \cup F_B)(x) = F_A(x) \cup F_B(x) = \langle 0 \wedge 1, 1 \vee 0\rangle = \langle 0,1\rangle$ and thus $x \in (F_A \cup F_B)^1$. Now $x \in A \cup B$. Therefore, $F_{A\cup B}(x) = \langle 0,1\rangle$ and hence $x \in (F_{A\cup B})^1$. Thus $(F_A \cup F_B)^1 = (F_A)^1 \cup (F_B)^1 = (F_{A\cup B})^1$.

	$x \in (F_A)^1$	$x \in (F_B)^1$	$x \in (F_A \cup F_B)^1$	$x \in (F_{A\cup B})^1$
(1)	T	F	T	T
(2)	F	T	T	T
(3)	T	T	T	T
(4)	F	F	F	F

	$x \in (F_A)^1$	$x \in (F_B)^1$	$x \in (F_A \cap F_B)^1$	$x \in (F_{A\cap B})^1$
(1)	T	F	F	F
(2)	F	T	F	F
(3)	T	T	T	T
(4)	F	F	F	F

	$x \in (F_A - F_B)^1$	$x \in (F_{A-B})^1$
$x \in A, x \notin B$	T	T
$x \notin A, x \in B$	F	F
$x \in A, x \in B$	F	F
$x \notin A, x \notin B$	F	F

Proof of result (4) follows from the following table. We distinguish four cases: (1) $x \in A, y \notin B$, (2) $x \notin A, y \in B$, (3) $x \in A, y \in B$ and (4) $x \notin A, y \notin B$.

	$x \in (F_A \times F_B)^1$	$x \in (F_{A\times B})^1$
$x \in A, y \notin B$	F	F
$x \notin A, y \in B$	F	F
$x \in A, y \in B$	T	T
$x \notin A, y \notin B$	F	F

To prove the result (5), we distinguish two cases: (1) $x \in A$ and (2) $x \notin A$. The proof follows from following table.

	$x \in (-F_A)^1$	$x \in (F_{A^c})^1$
$x \in A$	F	F
$x \notin A$	T	T

■

Theorem 1.3.13 *Let S be a set and let μ, σ be two fuzzy subsets on S.*

1. $\overline{F_\mu \cup F_\sigma} = \overline{F_\mu} \cup \overline{F_\sigma} = \overline{F_{\mu\cup\sigma}}$
2. $\overline{F_\mu \cap F_\sigma} = \overline{F_\mu} \cap \overline{F_\sigma} = \overline{F_{\mu\cap\sigma}}$
3. $\overline{F_\mu \times F_\sigma} = \overline{F_\mu} \times \overline{F_\sigma} = \overline{F_{\mu\times\sigma}}$

4. $\overline{-F_\mu} = \overline{F_{\mu^c}}$

Proof. Let $x \in S$. The proof is summarized in the following tables. Proof of first two results follows from the following two tables. We distinguish between two cases: (1) $\mu(x) \leq \sigma(x)$ and (2) $\mu(x) > \sigma(x)$.

As an illustration, we prove result (1) case (1) as follows: Assume that $\mu(x) \leq \sigma(x)$. Then $F_\mu(x) = \langle 1-\mu(x), \mu(x)\rangle$, $F_\sigma(x) = \langle 1-\sigma(x), \sigma(x)\rangle$. Hence $(\overline{F_\mu} \cup \overline{F_\sigma})(x) = \overline{F_\mu}(x) \cup \overline{F_\sigma}(x) = \mu(x) \vee \sigma(x) = \sigma(x)$. Now $(F_\mu \cup F_\sigma)(x) = F_\mu(x) \cup F_\sigma(x) = \langle (1-\mu(x)) \wedge (1-\sigma(x)), \mu(x) \vee \sigma(x)\rangle = \langle 1-\sigma(x), \sigma(x)\rangle$ and hence $\overline{F_\mu \cup F_\sigma}(x) = \sigma(x)$. Note also that $\underline{(\mu \cup \sigma)(x)} = \mu(x) \cup \sigma(x) = \sigma(x)$. Hence $F_{\mu\cup\sigma}(x) = \langle 1-\sigma(x), \sigma(x)\rangle$ and thus $\overline{F_{\mu\cup\sigma}}(x) = \sigma(x)$. Therefore, $\overline{F_\mu \cup F_\sigma} = \overline{F_\mu} \cup \overline{F_\sigma} = \overline{F_{\mu\cup\sigma}}$.

	$u(F_\mu(x))$	$u(F_\sigma(x))$	$u((F_\mu \cup F_\sigma)(x))$	$u(F_{\mu\cup\sigma}(x))$
$\mu(x) \leq \sigma(x)$	$\mu(x)$	$\sigma(x)$	$\sigma(x)$	$\sigma(x)$
$\mu(x) > \sigma(x)$	$\mu(x)$	$\sigma(x)$	$\mu(x)$	$\mu(x)$

	$u((F_\mu \cap F_\sigma)(x))$	$u(F_{\mu\cap\sigma}(x))$
$\mu(x) \leq \sigma(x)$	$\mu(x)$	$\mu(x)$
$\mu(x) > \sigma(x)$	$\sigma(x)$	$\sigma(x)$

Proof of result (3) follows from the following table. We distinguish between two cases: (1) $\mu(x) \leq \sigma(y)$ and (2) $\mu(x) > \sigma(y)$.

	$u((F_\mu \times F_\sigma)(x))$	$u(F_{\mu\times\sigma}(x))$
$\mu(x) \leq \sigma(y)$	$\mu(x)$	$\mu(x)$
$\mu(x) > \sigma(y)$	$\sigma(y)$	$\sigma(y)$

The proof of (4) follows from following table.

$u(-F_\mu(x))$	$u(F_{\mu^c}(x))$
$1-\mu(x)$	$1-\mu(x)$

■

Theorem 1.3.14 *Let S be a set and $A, B \subseteq S$. Then the following diagrams commute.*

$$\begin{array}{ccc} (A,B) & \xrightarrow{F} & (F_A, F_B) \\ \downarrow \cup & & \downarrow \cup \\ A \cup B = (F_A \cup F_B)^1 & \xleftarrow{(\,)^1} & F_A \cup F_B \end{array}$$

$$\begin{array}{ccc} (A,B) & \xrightarrow{F} & (F_A, F_B) \\ \downarrow \cap & & \downarrow \cap \\ A \cap B = (F_A \cap F_B)^1 & \xleftarrow{(\,)^1} & F_A \cap F_B \end{array}$$

$$\begin{array}{ccc} (A, B) & \xrightarrow{F} & (F_A, F_B) \\ \big\downarrow - & & \big\downarrow - \\ A - B = (F_A - F_B)^1 & \xleftarrow{(\,)^1} & F_A - F_B \end{array}$$

$$\begin{array}{ccc} (A, B) & \xrightarrow{F} & (F_A, F_B) \\ \big\downarrow \times & & \big\downarrow \times \\ A \times B = (F_A \times F_B)^1 & \xleftarrow{(\,)^1} & F_A \times F_B \end{array}$$

$$\begin{array}{ccc} A & \xrightarrow{F} & F_A \\ \big\downarrow c & & \big\downarrow - \\ A^c = (-F_A)^1 & \xleftarrow{(\,)^1} & -F_A \end{array}$$

Proof. Result follows from Theorem 1.3.12. ■

The Theorem 1.3.14 is a very powerful result. It states that given two sets A and B, the five operations union, intersection, difference, Cartesian product and complement, can either be computed as such or A and B can be converted into equivalent cisets and the corresponding operations can be carried out as cisets and converted back to sets without any loss of information. Since this result holds for all five operations, any operation defined using these five operations shall also inherit this property. This result clearly demonstrates the fact that the ciset is a generalization of the set.

Theorem 1.3.15 *Let S be a set and μ, σ be two fuzzy sets on S. Then the following diagrams commute.*

$$\begin{array}{ccc} (\mu, \sigma) & \xrightarrow{F} & (F_\mu, F_\sigma) \\ \big\downarrow \cup & & \big\downarrow \cup \\ \mu \cup \sigma = \overline{F_\mu \cup F_\sigma} & \xleftarrow{\overline{F}} & F_\mu \cup F_\sigma \end{array}$$

$$\begin{array}{ccc} (\mu, \sigma) & \xrightarrow{F} & (F_\mu, F_\sigma) \\ \big\downarrow \cap & & \big\downarrow \cap \\ \mu \cap \sigma = \overline{F_\mu \cap F_\sigma} & \xleftarrow{\overline{F}} & F_\mu \cap F_\sigma \end{array}$$

$$\begin{array}{ccc} (\mu,\sigma) & \xrightarrow{F} & (F_\mu, F_\sigma) \\ \downarrow \times & & \downarrow \times \\ \mu \times \sigma = \overline{F_\mu \times F_\sigma} & \xleftarrow{\overline{F}} & F_\mu \times F_\sigma \end{array}$$

$$\begin{array}{ccc} \mu & \xrightarrow{F} & F_\mu \\ \downarrow c & & \downarrow - \\ \mu^c = \overline{-F_\mu} & \xleftarrow{\overline{F}} & -F_\mu \end{array}$$

Proof. Result follows from Theorem 1.3.13 ■

The Theorem 1.3.15 is the fuzzy equivalent of the Theorem 1.3.14. In simple terms, it states that given two fuzzy subsets sets μ and σ, the four operations union, intersection, Cartesian product and complement can either be computed as fuzzy subsets or μ and σ can be converted into equivalent cisets and the corresponding operations can be carried out as cisets and then converted back to fuzzy sets without any loss of information. Since this result holds for all four operations, any operation defined using these operations shall also inherit this property. This result clearly demonstrates the fact that the ciset is a generalization of the fuzzy set.

In fact, from the above discussion, we have the following two results as well.

Theorem 1.3.16 *Let S be a set and $A, B \subseteq S$. Then the following diagrams commute.*

$$\begin{array}{ccc} (A,B) & \xrightarrow{F} & (F_A, F_B) \\ \downarrow \cup & & \downarrow \cup \\ A \cup B & \xrightarrow{F} & F_{A\cup B} = F_A \cup F_B \end{array}$$

$$\begin{array}{ccc} (A,B) & \xrightarrow{F} & (F_A, F_B) \\ \downarrow \cap & & \downarrow \cap \\ A \cap B & \xrightarrow{F} & F_{A\cap B} = F_A \cap F_B \end{array}$$

$$\begin{array}{ccc} (A,B) & \xrightarrow{F} & (F_A, F_B) \\ \downarrow - & & \downarrow - \\ A - B & \xrightarrow{F} & F_{A-B} = F_A - F_B \end{array}$$

$$\begin{array}{ccc} (A,B) & \xrightarrow{F} & (F_A,F_B) \\ \downarrow \times & & \downarrow \times \\ A\times B & \xrightarrow{F} & F_{A\times B}=F_A\times F_B \end{array}$$

$$\begin{array}{ccc} A & \xrightarrow{F} & F_A \\ \downarrow c & & \downarrow - \\ A^c & \xrightarrow{F} & F_{A^c}=-F_A \end{array}$$

■

Theorem 1.3.17 *Let S be a set and μ, σ be two fuzzy sets on S. Then the following diagrams commute.*

$$\begin{array}{ccc} (\mu,\sigma) & \xrightarrow{F} & (F_\mu,F_\sigma) \\ \downarrow \cup & & \downarrow \cup \\ \mu\cup\sigma & \xrightarrow{F} & F_{\mu\cup\sigma}=F_\mu\cup F_\sigma \end{array}$$

$$\begin{array}{ccc} (\mu,\sigma) & \xrightarrow{F} & (F_\mu,F_\sigma) \\ \downarrow \cap & & \downarrow \cap \\ \mu\cap\sigma & \xrightarrow{F} & F_{\mu\cap\sigma}=F_\mu\cap F_\sigma \end{array}$$

$$\begin{array}{ccc} (\mu,\sigma) & \xrightarrow{F} & (F_\mu,F_\sigma) \\ \downarrow \times & & \downarrow \times \\ \mu\times\sigma & \xrightarrow{F} & F_{\mu\times\sigma}=F_\mu\times F_\sigma \end{array}$$

$$\begin{array}{ccc} \mu & \xrightarrow{F} & F_\mu \\ \downarrow c & & \downarrow - \\ \mu^c & \xrightarrow{F} & F_{\mu^c}=-F_\mu \end{array}$$

■

In this section we have shown that ciset is in fact an extension of fuzzy sets. We can use this fact to introduce a difference operation on fuzzy sets. Of course, if $-$ denote the difference operation on fuzzy sets the following diagrams must commute:

$$\begin{array}{ccc} (\mu,\sigma) & \xrightarrow{F} & (F_\mu, F_\sigma) \\ \big\downarrow - & & \big\downarrow - \\ \mu - \sigma = \overline{F_\mu - F_\sigma} & \xleftarrow{\overline{F}} & F_\mu - F_\sigma \end{array}$$

$$\begin{array}{ccc} (\mu,\sigma) & \xrightarrow{F} & (F_\mu, F_\sigma) \\ \big\downarrow - & & \big\downarrow - \\ \mu - \sigma & \xrightarrow{F} & F_{\mu-\sigma} = F_\mu - F_\sigma \end{array}$$

Therefore, we define $(\mu - \sigma)(x) = \mu(x) \wedge (1 - \sigma(x))$.

1.4 Relations

Let S and T be two sets and let G and H be two cisets on S and T respectively. Then a subset $K \subseteq G \times H$ is called a *relation* from G to H. In other words, K is a ciset on $S \times T$ such that $K(x, y) \preceq G(x) \cap H(y)$, for all $(x, y) \in S \times T$. Thus the confidence index of a pair of elements never exceed the confidence index of either of the elements themselves. If we associate the elements as computers and pairs as the connecting communication links between the computers, the above restriction amounts to requiring that the confidence index of a communication link can never exceed the strengths of its connecting computers. The restriction $K(x, y) \subseteq G(x) \cap H(y)$, for all $(x, y) \in S \times T$ allows K_s^t to be a relation from G_s^t into H_s^t for all $\langle s, t\rangle \in \mathfrak{C}$.

There are three special cases of relations.

case 1: $S = T$ and $G = H$. In this case K is said to be a relation on G. Note that K is a subset of $G \times G$. Thus $K(x, y) \subseteq G(x) \cap G(y)$, for all $x, y \in S$.

case 2: $G(x) = \langle 0, 1\rangle$, for all $x \in S$ and $H(y) = \langle 0, 1\rangle$, for all $y \in T$. In this case, whenever there is no confusion, we may say K is a relation from S into T.

case 3: $S = T, G(x) = H(x) = \langle 0, 1\rangle$, for all $x \in S$. In this case, whenever there is no confusion, we may say K is a relation on S.

Example 1.4.1 *Let* $S = \{u, x, y, z\}$. *Define* $G(u) = \langle 0.4, 0.3\rangle$, $G(x) = \langle 0.3, 0.7\rangle, G(y) = \langle 0.2, 0.4\rangle$, $G(z) = \langle 0, 0.5\rangle$. *Clearly* G *is a ciset on* S. *Let* K *be ciset on* $S \times S$ *defined as follows:*

	u	x	y	z
u	$\langle 0.5, 0.2\rangle$	$\langle 0.7, 0.3\rangle$	$\langle 1, 0.2\rangle$	$\langle 0.6, 0.3\rangle$
x	$\langle 0.6, 0.3\rangle$	$\langle 0.4, 0.7\rangle$	$\langle 0.5, 0.2\rangle$	$\langle 0.8, 0.5\rangle$
y	$\langle 0.7, 0\rangle$	$\langle 0.1, 0.4\rangle$	$\langle 0.2, 0.3\rangle$	$\langle 1, 0.4\rangle$
z	$\langle 0.8, 0\rangle$	$\langle 0.9, 0\rangle$	$\langle 0.6, 0.2\rangle$	$\langle 0, 0.1\rangle$

Since $K(y,x) = \langle 0.1, 0.4\rangle \not\preceq \langle 0.3, 0.4\rangle = G(x) \cap G(y)$, K is not a relation on G. However, if we redefine $K(y,x) = \langle 0.3, 0.4\rangle$ then K is a relation on G.

Let K be a relation on S. Then K is called the strongest relation on G if for all relations J on $G, J \preceq K$, treating both J and K as subsets of $G \times G$. It is quite clear that K is the strongest relation on G, if and only if $K(x.y) = G(x) \cap K(y)$, for all $x, y \in S$.

Example 1.4.2 *Let $S = \{u, x, y, z\}$. Define $G(u) = \langle 0.9, 0.7\rangle$, $G(x) = \langle 0.5, 0.9\rangle, G(y) = \langle 0.6, 0.8\rangle$, $G(z) = \langle 0.4, 0.6\rangle$. Thus, G is a ciset on S. Let K be ciset on $S \times S$ be as follows:*

	u	x	y	z
u	$\langle 0.9, 0.7\rangle$	$\langle 0.9, 0.7\rangle$	$\langle 0.9, 0.7\rangle$	$\langle 0.9, 0.6\rangle$
x	$\langle 0.9, 0.7\rangle$	$\langle 0.5, 0.9\rangle$	$\langle 0.6, 0.8\rangle$	$\langle 0.5, 0.6\rangle$
y	$\langle 0.9, 0.7\rangle$	$\langle 0.6, 0.8\rangle$	$\langle 0.6, 0.8\rangle$	$\langle 0.6, 0.6\rangle$
z	$\langle 0.9, 0.6\rangle$	$\langle 0.5, 0.6\rangle$	$\langle 0.6, 0.6\rangle$	$\langle 0.4, 0.6\rangle$

It is easy to observe that K is the strongest relation on G.

The converse problem may also arise in practice. That is, we may know the confidence index of the pairs and we want to compute the minimum confidence index required for the elements themselves. For a given ciset K of $S \times S$, the weakest ciset G on S such that K is a relation on G is obtained by $G(x) = \cup\{K(x,y) \cup K(y,x) \mid y \in S\}$ for all $x \in S$.

Example 1.4.3 *Let $S = \{u, x, y, z\}$. Let K be ciset on $S \times S$ defined as follows:*

	u	x	y	z
u	$\langle 0.5, 0.2\rangle$	$\langle 0.2, 0.3\rangle$	$\langle 0.3, 0.2\rangle$	$\langle 0.1, 0.3\rangle$
x	$\langle 0.2, 0.3\rangle$	$\langle 0.1, 0.7\rangle$	$\langle 0.2, 0.2\rangle$	$\langle 0.3, 0.5\rangle$
y	$\langle 0.2, 0\rangle$	$\langle 0.2, 0.4\rangle$	$\langle 0.2, 0.3\rangle$	$\langle 0.5, 0.4\rangle$
z	$\langle 0.5, 0\rangle$	$\langle 0.4, 0\rangle$	$\langle 0.8, 0.2\rangle$	$\langle 0.9, 0.1\rangle$

Define $G : S \to \mathfrak{C}$ as below.
$S(u) = \langle 0.5, 0.2\rangle \cup \langle 0.2, 0.3\rangle \cup \langle 0.3, 0.2\rangle \cup \langle 0.1, 0.3\rangle \cup \langle 0.2, 0.3\rangle \cup \langle 0.2, 0\rangle \cup \langle 0.5, 0\rangle$,
$S(x) = \langle 0.2, 0.3\rangle \cup \langle 0.1, 0.7\rangle \cup \langle 0.2, 0.2\rangle \cup \langle 0.3, 0.5\rangle \cup \langle 0.2, 0.3\rangle \cup \langle 0.2, 0.4\rangle \cup \langle 0.4, 0\rangle$,

$S(y) = \langle 0,0\rangle \cup \langle 0.2,0.4\rangle \cup \langle 0.2,0.3\rangle \cup \langle 0.5,0.4\rangle \cup \langle 0.3,0.2\rangle \cup \langle 0.2,0.2\rangle \cup \langle 0.8,0.2\rangle$,
$S(z) = \langle 0.5,0\rangle \cup \langle 0,0\rangle \cup \langle 0,0.2\rangle \cup \langle 0.9,0.1\rangle \cup \langle 0.1,0.3\rangle \cup \langle 0.3,0.5\rangle \cup \langle 0.5,0.4\rangle$.
Thus $S(u) = \langle 0.1,0.3\rangle, S(x) = \langle 0.1,0.7\rangle, S(y) = \langle 0.2,0.4\rangle, S(z) = \langle 0.1,0.5\rangle$. *Clearly K is relation on G and G is the weakest ciset on S such that K is a relation on G.*

Given a relation K on a ciset G, it is possible to take (s,t)–cuts of both K and G to obtain a relation K_s^t on G_s^t.

Example 1.4.4 *Let $S = \{u,x,y,z\}$. Define $G(u) = \langle 0.4,0.3\rangle$, $G(x) = \langle 0.3,0.7\rangle, G(y) = \langle 0.2,0.4\rangle$, $G(z) = \langle 0,0.5\rangle$. Clearly G is a ciset on S. Let K be ciset on $S \times S$ defined as follows:*

	u	x	y	z
u	$\langle 0.5,0.2\rangle$	$\langle 0.7,0.3\rangle$	$\langle 1,0.2\rangle$	$\langle 0.6,0.3\rangle$
x	$\langle 0.6,0.3\rangle$	$\langle 0.4,0.7\rangle$	$\langle 0.5,0.4\rangle$	$\langle 0.8,0.5\rangle$
y	$\langle 0.7,0\rangle$	$\langle 0.1,0.4\rangle$	$\langle 0.2,0.4\rangle$	$\langle 1,0.4\rangle$
z	$\langle 0.8,0\rangle$	$\langle 0.9,0\rangle$	$\langle 0.6,0.2\rangle$	$\langle 0,0.5\rangle$

Choose $s = 0.5$ and $t = 0.4$. Then
$G_{0.5}^{0.4} = \{x,y,z\}$ *and* $K_{0.5}^{0.4} = \{(x,x),(x,y),(y,x),(y,y),(z,z)\}$.

Operations on relations

Let S be a set and G be a ciset on S. Given two relations J and K on G, J and K are cisets on $S \times S$. Therefore $J = K, J \prec K, J \preceq K, J \succ K, J \succeq K, J \cup K, J \cap K, -K$,and $J - K$ are defined. In particular, $J \cup K, J \cap K, -K, J - K$ are cisets on $S \times S$ and thus they are relations on G. For example, $J \cup K$ and $J \cap K$ are defined as follows: $(J \cup K)(x,y) = J(x,y) \cup K(x,y), (J \cap K)(x,y) = J(x,y) \cap K(x,y)$. Note that in particular, Theorem 1.3.6 hold.

We now proceed to introduce a new binary operation.

Definition 1.4.5 *Let S,T,U be sets and F,G,H be cisets on S,T and U respectively. Let K be relation from F to G and J be relation from G to H. Define $K \circ J : S \times U \to \mathfrak{C}$ by*

$$K \circ J(x,z) = \cup\{K(x,y) \cap J(y,z) \mid y \in T\}$$

for all $x \in S, z \in U$. The relation $K \circ J$ is called the composition *of K with J.*

Note that $K \circ J$ is a relation from ciset F to ciset H. A closer look at the definition of the composition operation reveals that $K \circ J$ can be computed similar to matrix multiplication, where the arithmetic operation addition is replaced by confidence index operation union and arithmetic operation multiplication is replaced by confidence index operation intersection.

Proposition 1.4.6 *The composition operation is associative.*

Note that in the special case, $S = T$ and $F = G, K \circ K$ is defined. Since composition is associative, we use the notation K^2 to denote $K \circ K$ and K^n to denote $K^{n-1} \circ K, n \succ 2$. Define $K^\infty(x, y) = \cup\{K^n(x, y) \mid n > 1\} \cup K(x, y)$. It may be recalled that we use notation K^1 to stand for the set $\{(x, y) \mid l(K(x, y)) = 0 \text{ and } u(K(x, y)) = 1\}$. Therefore, we shall avoid the notation K^1 to denote K. Further define K^0 by $K^0(x, y) = \langle 1, 0\rangle$ and $K^0(x, x) = F(x)$ otherwise.

Definition 1.4.7 *Let S, T be sets and F, G be cisets on S, T respectively. Let K be relation from F to G. Define $K^{-1} : T \times S \to \mathfrak{C}$ by*

$$K^{-1}(x, y) = K(y, x)$$

for all $x \in T, y \in S$. The relation $K^{-1}$is called the inverse *of K.*

We have the following result, proof of which is quite straight forward and hence we omit it.

Proposition 1.4.8 *Let S, T, U be sets and F, G, H be cisets on S, T, U respectively. Let K, K' be relations from F to G and J, J' be relation from G to H. Further, let $a = \langle s, t\rangle \in \mathfrak{C}$. Then*

1. $(K \circ J)^t_s \supseteq K^t_s \circ J^t_s$ and if S, T and U are finite, $(K \circ J)^t_s = K^t_s \circ J^t_s$
2. If $K \subseteq K'$ and $J \subseteq J'$ then $K \circ J \subseteq K' \circ J'$

Equivalence relations

Throughout this section, let S be a set and F be a ciset on S. Let K and J be two relations on F. It is quite natural to represent a relation in the form of a matrix. We now use the matrix representation of a relation to explain the properties of a relation. In particular, we shall use the term diagonal to represent the principal diagonal of the matrix.

We call K *reflexive* on F, if $K(x, x) = F(x)$ for all $x \in S$. If K is reflexive on F then $K(x, y) \preceq F(x) \cap F(y) \preceq F(x) = K(x, x)$ and it follows that any diagonal element is larger than or equal to any element in its row. Similarly, any diagonal element is larger than or equal to any element in its column. Conversely, given a relation K on F such that any diagonal element is larger than or equal to any element in its row and column, define a ciset G on S as $G(x) = K(x, x)$, for all $x \in S$. Then G is the weakest ciset on S such that K is a relation on G. Further, K is reflexive on G.

Reflexive relations have some interesting algebraic properties.

Theorem 1.4.9 *Let S be a set and F be a ciset on S and let K and J be two relations on F. Then the following properties hold.*

1. If K is reflexive, $J \subseteq J \circ K$ and $J \subseteq K \circ J$,
2. If K is reflexive, $K^0(x,x) = K(x,x) = K^2(x,x) = \ldots = K^\infty(x,x) = F(x), \forall x \in S$,
3. If K is reflexive, $K^0 \subseteq K \subseteq K^2 \subseteq \ldots \subseteq K^\infty$,
4. If K and J are reflexive, so is $J \circ K$ and $K \circ J$,
5. If K is reflexive, then K_s^t is a reflexive relation on F_s^t for all confidence indexes $\langle s,t \rangle$.

Proof. Let $x, z \in S$.

1. $(K \circ J)(x,z) = \cup\{K(x,y) \cap J(y,z) \mid y \in S\} \succeq K(x,x) \cap J(x,z) = F(x) \cap J(x,z)$. Recall that $J(x,z) \preceq F(x) \cap F(z)$. Therefore, $F(x) \cap J(x,z) = J(x,z)$. Thus $J \subseteq J \circ K$. The result, $J \subseteq K \circ J$ follows similarly.
2. Note that $K(x,x) = F(x), \forall x \in S$. Assume that $K^n(x,x) = F(x)$, $\forall x \in S$. Now for all $x \in S$, $K^{n+1}(x,x) = \cup\{K(x,y) \cap K^n(y,x) \mid y \in S\} \preceq \cup\{F(x) \cap F(x) \mid y \in S\} = F(x)$ and $K^{n+1}(x,x) = \cup\{K(x,y) \cap K^n(y,x) \mid y \in S\} \succeq K(x,x) \cap K^n(x,x) = F(x)$. Hence the result.
3. Choose J to be the same as K in (1). Thus we get $K \subseteq K^2$. By (2), K^2 is reflexive. Now apply (1) by choosing J as K^2. Thus $K^2 \subseteq K^3$. Repeating this process, the result follows.
4. $(K \circ J)(x,x) = \cup\{K(x,y) \cap J(y,x) \mid y \in S\} \preceq \cup\{F(x) \cap F(x) \mid y \in S\} = F(x)$ and $(K \circ J)(x,x) = \cup\{K(x,y) \cap J(y,x) \mid y \in S\} \succeq K(x,x) \cap J(x,x) = F(x) \cap F(x) = F(x)$. The proof that $K \circ J$ is reflexive is similar.
5. If $x \in F_s^t$, then $\langle s,t \rangle \preceq F(x) = K(x,x)$. Therefore $(x,x) \in K_s^t$.

■

We call K *symmetric* if $K(x,y) = K(y,x)$, for all $x, y \in S$. In other words, K is symmetric if the matrix representation of K is symmetric.

Theorem 1.4.10 *Let S be a set and F be a ciset on S and let K and J be two relations on F. Then the following properties hold.*

1. If K and J are symmetric then $J \circ K$ is symmetric if and only if $J \circ K = K \circ J$,
2. If K is symmetric, then so is every power of K,

3. If K is symmetric, then K_s^t is a symmetric relation on F_s^t for all confidence indexes $\langle s, t \rangle$.

Proof. Let $x, z \in S$.

1. $(K \circ J)(x, z) = (K \circ J)(z, x)$
   $\Leftrightarrow \cup\{K(x, y) \cap J(y, z) \mid y \in S\} = \cup\{K(z, y) \cap J(y, x) \mid y \in S\}$
   $\Leftrightarrow \cup\{K(x, y) \cap J(y, z) \mid y \in S\} = \cup\{J(x, y) \cap K(y, z) \mid y \in S\}$
   $\Leftrightarrow K \circ J = J \circ K$

2. Assume that K^n is symmetric for $n \in \mathbb{N}$. Then $K^{n+1}(x, z)$
   $= \cup\{K(x, y) \cap K^n(y, z) \mid y \in S\}$
   $= \cup\{K(y, x) \cap K^n(z, y) \mid y \in S\}$
   $= \cup\{K^n(z, y) \cap K(y, x) \mid y \in S\}$
   $= K^{n+1}(z, x)$. Hence the result.

3. If $(x, z) \in K_s^t$ then $\langle s, t \rangle \preceq K(x, z) = K(z, x)$. Therefore $(z, x) \in K_s^t$.

■

We call K transitive if $K^2 \subseteq K$. It follows that K^∞ is always transitive for any relation K.

Theorem 1.4.11 *Let S be set and F be cisets on S and let K, J and L be three relations on F. Then the following properties hold.*

1. If K is transitive and $J \subseteq K, L \subseteq K$ then $J \circ L \subseteq K$,

2. If K is transitive, then so is every power of K,

3. If K is transitive, J is reflexive and $J \subseteq K$ then $K \circ J = J \circ K = K$,

4. If K is reflexive and transitive then $K^0 \subseteq K = K^2 = \ldots = K^\infty$,

5. If K and J are transitive and $J \circ K = K \circ J$, then $K \circ J$ is transitive,

6. If K is symmetric and transitive then $K(x, y) \preceq K(x, x)$ and $K(y, x) \preceq K(x, x)$, for all $x, y \in S$,

7. If K is transitive, then K_s^t is a reflexive relation on F_s^t for all confidence indexes $\langle s, t \rangle$.

Proof. Let $x, z, w \in S$.

1. $(J \circ L)(x, z) = \cup\{J(x, y) \cap L(y, z) \mid y \in S\} \preceq \cup\{K(x, y) \cap K(y, z) \mid y \in S\} = K^2(x, z) \preceq K(x, z)$.

2. Assume that K^n is transitive. Then $K^n \circ K^n \subseteq K^n$. Now $K^{n+1} \circ K^{n+1} = K^{2n} \circ K^2 \subseteq K^n \circ K = K^{n+1}$.

3. By (1), $K \circ J \subseteq K$. Note that $(K \circ J)(x,z) = \cup\{K(x,y) \cap J(y,z) \mid y \in S\} \succeq K(x,z) \cap J(z,z) = K(x,z) \cap F(z) = K(x,z)$. Hence $K \circ J = K$. Similarly, $J \circ K = K$.

4. By (3), $K \circ K = K$. Thus $K^2 = K$. Assume that $K^n = K$. Since K^n is transitive and K is reflexive, by (3), $K^n \circ K = K$. Thus $K^{n+1} = K$.

5. $(K \circ J) \circ (K \circ J) = K \circ (J \circ K) \circ J = K \circ (K \circ J) \circ J \subseteq K \circ J$. Thus $K \circ J$ is transitive.

6. Since K is transitive, $K \circ K \subseteq K$. Hence $(K \circ K)(x,x) \preceq K(x,x)$. That is, $\cup\{K(x,y) \cap K(y,x) \mid y \in S\} \preceq K(x,x)$. Note that K is symmetric and hence $K(x,y) = K(y,x)$. Thus $\cup\{K(x,y) \cap K(x,y) \mid y \in S\} \preceq K(x,x)$ and hence $K(x,y) \preceq K(x,x)$. Once again, K being symmetric, $K(y,x) \preceq K(x,x)$.

7. Let $(x,w), (w,z) \in K_s^t$. Hence $K(x,w) \succeq \langle s,t \rangle$ and $K(w,z) \succeq \langle s,t \rangle$. Therefore, $K(x,w) = \cup\{K(x,y) \cap K(y,z) \mid y \in S\} \succeq K(x,w) \cap K(w,z) \succeq \langle s,t \rangle$. Thus $(x,w) \in K_s^t$.

■

A relation K on a ciset H that is reflexive, symmetric, and transitive is called an equivalent relation on H.

In this section we have proved many the results on ciset relations. All these results and definitions have a corresponding fuzzy equivalent. In fact, if you have a result in fuzzy sets, the corresponding result in ciset also holds. However, a ciset can not be considered as a fuzzy set due to the fact that complement operation as defined in this chapter is unique to ciset and can not be simulated through fuzzy set theory. Further, ciset can be used to introduce a difference operation in fuzzy sets.

2
THE RELATIONAL MODEL

In this chapter we introduce the relational model, a collection of methods and techniques for organizing databases centered on the notion of a relation as the data structure. After we present the notion of a relation, we present other fundamental concepts such as candidate key, primary key and foreign key. Integrity rules such as the entity integrity rule and the referential integrity rule are presented. We conclude this chapter with a brief introduction to relational operations.

A model is a representation of some real world object or idea. For example, a house can be modeled using blueprints or a car can be modeled as a small toy car. Before one builds a house, it is customary to draw a blueprint of the house. The blueprint is drawn using a set of well-understood symbols and it forms the basis of the contract between the builder and the owner of the house. Further, it is quite easy to make changes without tearing down walls. However, a blueprint is just a visual representation of the house and it does not have all the properties of a house. In general, a model helps us understand a real world object or an idea. It further allows us to make changes and uses commonly agreed upon symbols to convey the idea between the builder and the user of a system.

A relational model is a blueprint of a database. In a relational model, the data is presented in the form of a table, which we call a relation. Every piece of information is stored as one or more relations. Informally, a relational database can be conceived of as a collection of relations. For example, the employee information can be organized in the form of as a relation as shown in Table 2.1.

TABLE 2.1 EMPLOYEE

EMPID	EMPNAME	DEPT	PHONE	RMNUM
12312	John Smith	Marketing	212-3456	MK211
31897	Mary Lee	Services	345-2647	AD345
12674	Sandy Dewitt	Marketing	339-7595	MK212
56739	Bea Anthony	Accounting	234-3498	AD438
42677	Cathy Li	Accounting	452-3462	AD425
64828	Joy Austin	Services	899-4643	AD324
42816	Chris Kroll	Accounting	525-3462	AD437
87342	Matt Ridle	Services	233-2445	AD346
95363	Jack Wall	Marketing	341-3577	MK213
72358	James Wood	Marketing	341-3389	MK215

As we contemplate the EMPLOYEE table, we can identify several elements.

- the name of the table: EMPLOYEE,
- the column headings of the table: EMPID, EMPNAME, DEPT, PHONE, RMNUM,
- the set of rows of the table.

The column headings of table are known as relational *attributes*. Thus the EMPLOYEE table has five attributes: EMPID, EMPNAME, DEPT, PHONE and RMNUM. Each row of the table is known as a tuple. Thus the above table has ten tuples.

2.1 Formalization of Relation

We now proceed to formalize the notion of a table. Let U be the set of all relational attributes. For each attribute $A \in U$, let $DOM(A)$, called domain of A, denote the set of all possible values that can occur in that column. It is safe to assume that $DOM(A)$ contains at least two different values. The domains are arbitrary, nonempty sets that can be either finite or countably infinite. Note that if there is only one element in $DOM(A)$, we need not include A as an attribute of the table.

Let $R = \{A_1, \ldots, A_n\}$ be a finite set of relational attributes. Then R is called a *relational scheme*. A *relation* r on relational scheme R is a finite set of mappings $\{t_1, \ldots, t_m\}$ from $\cup\{DOM(A_i) \mid i = 1, \ldots, n.\}$ with the restriction that $t(A_i) \in DOM(A_i), i = 1, \ldots, n$ for $t \in \{t_1, \ldots, t_m\}$. The mappings are called tuples.

Example 2.1.1 *In Table 2.1, the relational scheme is*
$EMPLOYEE\_SCHEME$
$= \{EMPID, EMPNAME, DEPT, PHONE, RMNUM\}$.
The tuple
$t = (12312, John\ Smith, Marketing, 212 - 3456, MK211)$
can be considered as a mapping defined on the set
$\{EMPID, EMPNAME, DEPT, PHONE, RMNUM\}$
by
$t(EMPID) = 12312,$
$t(EMPNAME) = John\ Smith,$
$t(DEPT) = Marketing,$
$t(PHONE) = 212 - 3456,$
$t(RMNUM) = MK211.$
Thus the EMPLOYEE relation has ten tuples and five attributes. The domains for each of these attributes are as follows.
$DOM(EMPID)$ = *set of all five digit numbers,*
$DOM(EMPNAME)$ = *set of all names,*
$DOM(DEPT)$ = *{Marketing, Services, Accounting},*
$DOM(PHONE)$ = *the set of phone numbers,*
$DOM(RMNUM)$ = *the set of room numbers.*

Properties of a relation

Property 1: The order of attributes is immaterial

One important consequence of considering tuples as mappings is that the order of attributes is immaterial from a theoretical point of view. We now proceed to show that the order of attributes is immaterial from a practical point of view. For example, the following two relations are the same.

Consider two relations EMPLOYEE and EMPLYE on the relational scheme EMPLOYEE_SCHEME given below.

EMPLOYEE

EMPID	EMPNAME	DEPT	PHONE	RMNUM
12312	John Smith	Marketing	212-3456	MK211
31897	Mary Lee	Services	345-2647	AD345
12674	Sandy Dewitt	Marketing	339-7595	MK212
56739	Bea Anthony	Accounting	234-3498	AD438

EMPLYE

RMNUM	PHONE	EMPNAME	DEPT	EMPID
MK211	212-3456	John Smith	Marketing	12312
AD345	345-2647	Mary Lee	Services	31897
MK212	339-7595	Sandy Dewitt	Marketing	12674
AD438	234-3498	Bea Anthony	Accounting	56739

Note that the tuple 1 in the EMPLOYEE relation and the tuple 1 in the EMPLYE relation have the same information content that can be stated in plain English as "John Smith is in Marketing department, has employee identification number 12312, phone number 212-3456 and his room number is MK211". The second tuple in the EMPLOYEE relation and the second tuple in the EMPLYE relation conveys the fact that Mary Lee is Customer service has employee identification number 31897, phone number 345-2647 and her room number is AD345 and so on. Thus each tuple conveys the same semantic interpretation in both relations and hence both relations are identical in all respects from the practical perspective.

Property 2: The order of tuples in a relation is also immaterial

A relation is considered a set of mappings. Therefore, the order of tuples in a relation is also immaterial from a theoretical point of view. We now proceed to show that the order of tuples in a relation is immaterial from a practical point of view. For example, the following two relations are the same.

Consider two relations EMPLOYEE and EMP on the relational scheme EMPLOYEE_SCHEME given below.

EMPLOYEE

EMPID	EMPNAME	DEPT	PHONE	RMNUM
12312	John Smith	Marketing	212-3456	MK211
31897	Mary Lee	Services	345-2647	AD345
12674	Sandy Dewitt	Marketing	339-7595	MK212
56739	Bea Anthony	Accounting	234-3498	AD438

EMP

EMPID	EMPNAME	DEPT	PHONE	RMNUM
31897	Mary Lee	Services	345-2647	AD345
56739	Bea Anthony	Accounting	234-3498	AD438
12312	John Smith	Marketing	212-3456	MK211
12674	Sandy Dewitt	Marketing	339-7595	MK212

The tuple 1 of the EMPLOYEE relation appears as the tuple 3 of the EMP relation. The information content of these two tuples is the same and can be stated in plain English as "John Smith is in Marketing department, has employee identification number 12312, phone number 212-3456 and his room number is MK211". Similarly, the tuple 2 of the EMPLOYEE relation appears as the tuple 4 of the EMP relation and so on. Thus every tuple of the EMPLOYEE relation appears as a tuple of the EMP relation and vice-versa, except for their order of appearances. Since a tuple conveys the same semantic interpretation in both relations EMPLOYEE and EMP are identical in all respects from the practical perspective.

Property 3: No two tuples are identical in a relation

A relation is considered a set of mappings and each mapping corresponds to a tuple in the relation. From the definition of a set, no two members of a set are identical. Therefore, no two mappings in a relation are identical. In other words, no two tuples in a relation are identical.

By keeping two identical tuples, we are simply restating the same information. Therefore, from a practical point of view, there is no need to keep identical tuples in a relation.

Property 4: A relation is in first normal form

A relation is considered a set of mappings. Each mapping corresponds to one tuple in the relation. From the definition of relation, each mapping assigns a single value to each of the attributes. This property is known as first normal form. This being an important property, we devote the next subsection to further illustrate it.

First normal form

A relation is in *First Normal Form (1NF)*, if every attribute value is an atomic value and not a list or a set of values. We always keep a relation in first normal form. That is, no tuple in the relation has more than one value for each of its attributes. Note that this requirement is implicit in our definition of a relation.

Example 2.1.2 *Following is a table of employees and cities they have lived before.*

LIVED

EMPNAME	*CITY*
John Smith	*New York, Chicago, Los Angeles*
Mary Lee	*New York, San Francisco*
Sandy Dewitt	*Boston, Paris, New Delhi*
Bea Anthony	*Kansas City*

This table is not in first normal form. In the case of John Smith, the CITY attribute has multiple values. So is the case with Mary Lee and Sandy Dewitt. In other words, the DOM(CITY) = {{ New York, Chicago, Los Angeles}, { New York, San Francisco}, { Boston, Paris, New Delhi}, { Kansas City}}. Thus elements of DOM(CITY) are sets; and not atomic values. We can convert this table to enforce the first normal form constraint as follows.

LIVED

EMPNAME	*CITY*
John Smith	*New York*
John Smith	*Chicago*
John Smith	*Los Angeles*
Mary Lee	*New York*
Mary Lee	*San Francisco*
Sandy Dewitt	*Boston*
Sandy Dewitt	*Paris*
Sandy Dewitt	*New Delhi*
Bea Anthony	*Kansas City*

2.2 Integrity Constraints

superkey

Consider the Table 2.1. In the relation EMPLOYEE, the attribute EMPID stands for employee identification number. This number is unique for each employee of a particular company. No two employees can have the same employee identification number. We call such an attribute a superkey.

A superkey is an attribute or a combination of attributes that uniquely identifies each tuple of a relation.

The attribute EMPNAME is not unique to each of the employees. Even though at this time, there is only one John Smith among the employees, there is no reason to believe that in the future there will not be another employee with the same name. Thus EMPNAME can not be a superkey. The remaining attributes, DEPT, PHONE, or RMNUM can not be superkeys either.

To illustrate the concept of a superkey, let us consider the relation FOUNDER presented in Table 2.2.

TABLE 2.2 FOUNDER

FNAME	LNAME	SSN	ACODE	PHONE
Mary	Shue	131-21-4567	402	2345678
Mary	Buckett	456-58-4557	511	2327890
Joy	Buckett	256-67-3253	511	3414567
Joy	Wright	679-77-4512	345	2345678

For the sake of discussion, let us assume that the table has only four tuples and these are the only tuples ever going to be in the table. Let us consider the attribute FNAME. Since the value Mary appears more than once, the attribute FNAME can not be used to uniquely identify a tuple. So FNAME can not be a superkey. Since the value Buckett appears twice, the attribute LNAME can not be a superkey. However, if we consider two attributes FNAME and LNAME together, each tuple can be uniquely identified. Therefore, the pair of attributes {FNAME,LNAME} is a superkey. SSN is a superkey, since the social security number is a unique to each person. In this example, ACODE stands for the area code. Note that Mary Buckett and Joy Buckett share the same area code. Thus ACODE can not be a superkey. Since the PHONE value 2345678 is the same for Mary Shue and Joy Wright, PHONE can not be a superkey. It is worth noticing that ACODE along with PHONE will uniquely identify a founder and hence those two attributes together form a superkey.

Formalization of superkey

Let $r = \{t_1, \ldots, t_m\}$ be a relation on relational scheme $R = \{A_1, \ldots, A_n\}$. A nonempty subset of relational attributes X of R is called a superkey, if for any two distinct tuples $t, t' \in r, t(X)$ is different from $t'(X)$.

Consider the relation FOUNDER in Table 2.2. The relational scheme is given by

FOUNDER_SCHEME = {FNAME, LNAME, SSN, ACODE, PHONE}.

The relation has four tuples $t_1, \ldots, t_4$, where

$t_1 = (Mary, Shue, 131 - 21 - 4567, 402, 2345678)$,
$t_2 = (Mary, Buckett, 256 - 67 - 3253, 511, 2327890)$,
$t_3 = (Joy, Buckett, 456 - 58 - 4557, 511, 3414567)$,
$t_4 = (Joy, Wright, 679 - 77 - 4512, 345, 2345678)$,

- Let $X = \{FNAME\}$. Note that $t_1(X) = (Mary) = t_2(X)$. Therefore $\{FNAME\}$ is not a superkey.

- Let $X = \{LNAME\}$. Now, $t_2(X) = (Buckett) = t_3(X)$. Thus $\{LNAME\}$ is not a superkey.

- Let $X = \{FNAME, LNAME\}$. It can be seen that $t_1(X) = (Mary, Shue), t_2(X) = (Mary, Buckett), t_3(X) = (Joy, Buckett)$ and $t_4(X) = (Joy, Wright)$. Note that all these pairs of values are unique, and we further know that these are the only possible values for the pair of attributes FNAME and LNAME. Thus the condition that "for any two distinct tuples t, t' of FOUNDER, $t(X)$ is different from $t'(X)$" is met and thus $\{FNAME, LNAME\}$ is a superkey. It must be noted that it is not a good idea, in general, to treat $\{FNAME, LNAME\}$ as superkey. No one can really guarantee that there won't be two people with same first and last names. Our aim in this example is to illustrate the concept involved using a simple and familiar context.

- Let $X = \{SSN\}$. It is a well-known fact that social security number is unique for each one of us. Therefore, the condition "for any two distinct tuples t, t' of FOUNDER, $t(X)$ is different from $t'(X)$" is met and thus $\{SSN\}$ is a superkey.

Properties of a superkey

Property 1: Every relation has a superkey

This is an immediate consequence of the Property 3 of a relation. According to Property 3 of a relation, no two tuples of a relation are identical. Therefore, if r is a relation on relational scheme $R, t(R)$ and $t'(R)$ are not identical for any two tuples t, t' of r. Thus R is a superkey.

Example 2.2.1 *Consider the relation FOUNDER presented in Table 2.2. Choose X to be the same as R. That is,*
$X = \{FNAME, LNAME, SSN, ACODE, PHONE\}$. *Now,*
$t_1(X) = t_1 = (Mary, Shue, 131-21-4567, 402, 2345678)$,
$t_2(X) = t_2 = (Mary, Buckett, 256-67-3253, 511, 2327890)$,
$t_3(X) = t_3 = (Joy, Buckett, 456-58-4557, 511, 3414567)$,
$t_4(X) = t_4 = (Joy, Wright, 679-77-4512, 345, 2345678)$.
Clearly, none of the tuples t_1, t_2, t_3 and t_4 is identical. Therefore
$X = \{FNAME, LNAME, SSN, ACODE, PHONE\}$
is a superkey.

Property 2: If X is a superkey, so is any super set of X.

This property needs to be explained in more detail. Let r be a relation on a relational scheme R and let X be a superkey of r. The Property 2 may be stated in more detail as follows:

Let X be a superkey on relational scheme R and Y be such that $X \subseteq Y \subseteq R$. Then Y is a superkey.

Assume that X is a superkey. Therefore, for any two tuples t, t' of $r, t(X)$ and $t'(X)$ are not identical. Now Y has all the attributes of X and as a consequence for any two tuples t, t' of $r, t(Y)$ and $t'(Y)$ are not identical. Thus Y is a superkey.

Example 2.2.2 *Consider the relation FOUNDER presented in Table 2.2. We noticed that $X = \{SSN\}$ is a superkey. Let $Y = \{FNAME, SSN\}$. Note that $S = \{SSN\} \subseteq Y = \{FNAME, SSN\} \subseteq R$*
$= \{FNAME, LNAME, SSN, ACODE, PHONE\}$. *Property 2 states that $\{FNAME, SSN\}$ is a superkey. This can be verified by observing the fact that all of the following tuples*
$t_1(Y) = (Mary, 131-21-4567)$,
$t_2(Y) = (Mary, 256-67-3253)$,
$t_3(Y) = (Joy, 456-58-4557)$,
$t_4(Y) = (Joy, 679-77-4512)$,
are distinct.

Property 3: Every superkey contains a minimal superkey.

The term minimal warrants some explanation. Starting with a superkey X, it is possible to remove one of relational attributes from X to obtain a proper subset Y of X. Check whether or not Y is a superkey. If Y is not a superkey, we ignore Y and start with X again. However, if Y is a superkey, we replace X with Y and repeat the process again. This process will end with a set X of relational attributes such that X is a superkey and no proper subset of X is a superkey. The superkey X obtained by the above process is called a minimal superkey.

A superkey X is a minimal superkey if no proper subset of X is a superkey.

In Example 2.2.1, we have seen that
$X = \{FNAME, LNAME, SSN, ACODE, PHONE\}$
is a superkey. Let us remove PHONE from X to obtain
$Y = \{FNAME, LNAME, SSN, ACODE\}$.
It is left to the reader to verify that Y is a superkey. Therefore, we rename Y as X. We now repeat the above steps with new X.
$X = \{FNAME, LNAME, SSN, ACODE\}$
is a superkey. Let us remove ACODE from X. Now
$Y = \{FNAME, LNAME, SSN\}$. Once again, reader is urged to verify the fact that Y is a superkey.

$X = \{FNAME, LNAME, SSN\}$ is a superkey. Remove SSN from X. Now $Y = \{FNAME, LNAME\}$ is a superkey.

$X = \{FNAME, LNAME\}$ is a superkey. Remove LNAME from X. Now $Y = \{FNAME\}$. We have observed in Example 2.2.1 that Y is not a superkey. Therefore, we ignore Y.

$X = \{FNAME, LNAME\}$ is a superkey. Remove FNAME from X. Now $Y = \{LNAME\}$ and Y is not a superkey. Therefore, we ignore Y.

Thus from X, we cannot remove any attribute to obtain a proper subset of X which is a superkey. Therefore, X is a minimal superkey.

Example 2.2.3 *Continuing with Example 2.2.1,*
$X = \{FNAME, LNAME, SSN, ACODE, PHONE\}$
is a superkey. Remove FNAME from X to obtain
$Y = \{LNAME, SSN, ACODE, PHONE\}$
and note that Y is a superkey. Therefore, we rename Y as X.
$X = \{LNAME, SSN, ACODE, PHONE\}$ *is a superkey. Let us remove ACODE from X. Now $Y = \{LNAME, SSN, PHONE\}$ and Y is a superkey.*
$X = \{LNAME, SSN, PHONE\}$ *is a superkey. Remove LNAME from X. Now $Y = \{SSN, PHONE\}$ is a superkey.*
$X = \{SSN, PHONE\}$ *is a superkey. Remove PHONE from X. Now $Y = \{SSN\}$ and Y is a superkey. So we rename Y as X.*

$X = \{SSN\}$ is a superkey. Removal of SSN from X leads to an empty set. Therefore, X is a minimal superkey.

Example 2.2.4 *Consider the relation FOUNDER presented in Table 2.2. We have seen that*
$\{FNAME, LNAME\}$ is a minimal superkey. Further, $\{SSN\}$ is also a minimal superkey. Thus a relational may have many minimal superkeys.

Candidate key

In the previous subsection, we have noticed that a relation may have many superkeys. In fact, we have seen that a relation may have many minimal superkeys. Minimal superkeys play an important role in database design and they are called candidate keys.

A *candidate key* is an attribute or a combination of attributes that uniquely identifies each tuple of a relation and is minimal.

Let $r = \{t_1, \ldots, t_m\}$ be a relation on relational scheme $R = \{A_1, \ldots, A_n\}$. A nonempty subset of relational attributes X of R is called a candidate key, if for any two distinct tuples $t, t' \in r, t(X)$ is different from $t'(X)$ and for every proper subset of X' of X, there exists at least two distinct tuples $t, t' \in r, t(X')$ is the same as $t'(X')$.

In terms of superkeys, a candidate key can be formally defined as follows.

Let $r = \{t_1, \ldots, t_m\}$ be a relation on relational scheme $R = \{A_1, \ldots, A_n\}$. A nonempty subset of relational attributes X of R is called a candidate key, if X is a superkey and no proper subset of X is a superkey.

Example 2.2.5 *Consider Example 2.2.1. The attribute SSN being social security number, SSN is unique for each of the individuals. Thus SSN is a superkey. Since SSN is a single attribute, SSN is a minimal superkey.*

Example 2.2.6 *Consider the relational schema given below.*

BOOK = {ISBN, LCCN, AU_ONE, AU_TWO, TITLE, EDITION, VOL, PUBL}

Where
ISBN : International Standard Book Number,
LCCN: Library of Congress Catalog Number,
AU_ONE: First Author,
AU_TWO: Second Author,
TITLE: Title of the book,
EDITION: Edition number,
VOL: Volume number,
PUBL: Publisher.

ISBN is unique to each book. Therefore, ISBN is a superkey. Further, ISBN is a single attribute, it is a minimal superkey. Thus ISBN is a candidate key. LCCN is also unique to each book and LCCN is a single attribute. Thus LCCN is a candidate key.

Example 2.2.7 *An employee of a company may have an employee number that is a unique identifier. Further, an employee has a social security number. Thus a relational schema containing two attributes EMPNO (employee number) and SSN (social security number) has two candidate keys: EMPNO and SSN.*

Example 2.2.8 *Every vehicle has a unique identification number called VIN (vehicle identification number). Clearly VIN is a candidate key. In United States, each state issues a unique license plate number. Therefore, state along with the license plate number is also a candidate key.*

Primary key

In previous subsection, we have observed that a relation may have many candidate keys. One of the candidate keys is selected as the primary key of the relation. It is important to consider various issues in the selection process.

Ownership and privacy: The idea may be better explained through an example. Consider the Example 2.2.7. The company owns the employee number. On the other hand, social security number is used by many organizations and as such an employee may want to protect his or her social security number. Thus, in this context, employee number is to be chosen as the primary key.

Convenience and information sharing: Example 2.2.8 illustrates the concept involved. Note that VIN is not determined by any insurance company or any state motor vehicle department. However, it is better to use VIN as the primary key due to the fact that privacy is not an issue in this case; and information sharing is more important.

Once a primary key is selected, we insist that primary key value cannot be null. That is, all attributes involved in the primary key must have a valid data entry. From the definition of the primary key, it follows that primary key value cannot be repeated. Thus we have the *first data integrity rule*, also known as, *entity integrity rule.*

A primary key value cannot be repeated and it cannot be a null value.

The purpose of this integrity rule is to ensure that every tuple in a relation has a unique identifier.

Every relation in a database must obey the entity integrity rule.

Foreign key

Another level of data integrity is enforced through foreign keys. Consider the two relations AGENT and CUSTOMER maintained by an insurance company.

TABLE 2.3 AGENT

FNAME	LNAME	AGNT_NBR	AGNT_ADRS	PHONE
Molly	Smith	1	12767 T St.	4444568
Mike	Covey	2	67829 Q Ave.	5785678
Sue	Newman	3	462 S 107 St.	6489490
Chris	Levy	4	102 N 38 St.	3451919

TABLE 2.4 CUSTOMER

FNAME	LNAME	CNBR	CUST_ADRS	AMT	AGNT
John	Shaw	10001	16767 J Ave.	100,000	1
Ravi	Ray	10002	629 A St.	500,000	2
Magan	Fields	10003	568 N 17 St.	200,00	1
Joyce	Lendl	10004	88 S 87 St.	750,000	

In this example, the AGENT table keeps all the pertinent information on different agents. For the sake of convenience, a unique number called "AGNT_NBR" is assigned to each of the agents and this number will serve as the primary key of the AGENT relation. The relation CUSTOMER on the other hand, keeps the information about different customers. It is common practice that each customer is assigned to at most one agent. However, an agent may have many customers. Thus there is a relationship between the customers and agents and the attribute AGNT in the CUSTOMER table is used to keep track of this relationship.

What are the possible values that can appear as AGNT in the CUSTOMER relation? In other words, what is the domain of the attribute AGNT? Clearly, it is the set of all agent numbers. Sometimes an agent may not be assigned to a customer as in the case of customer Joyce Lendl. In that case, the attribute AGNT has null value. Once an agent has been assigned to a particular customer, the assigned agent's AGNT_NBR is the only possible value for the attribute AGNT of that particular customer. Thus possible values for the attribute AGNT are all the values in the attribute AGNT_NBR of the relation AGENT along with the null value. In this case, we call the attribute AGNT of the relation CUSTOMER a foreign key referencing AGENT relation.

An attribute F of a relation is a foreign key if $DOM(F)$ is a subset of $DOM(P)$, where P is a primary key of some relation in the database.

Example 2.2.9 *Consider the relation AGENT in Table 2.3. The primary key of this relation is AGNT_NBR. Note that DOM(AGNT_NBR) is the set of all agent numbers. Now the domain of AGNT in relation CUSTOMER (Table 2.4) denoted by DOM(AGNT) is the set of all agent numbers. Note that domain is the set of all possible values and as such, in this case, the domain of AGNT is the set of all agent numbers and not just {1,2}. Thus, DOM(AGNT) is a subset of DOM(AGNT_NBR) and*

AGNT_NBR is a primary key of a relation AGENT in the database. Therefore, AGNT is a foreign key.

Example 2.2.10 *Consider the relation CRS given below.*

CRS

CRS_NBR	*CRS_TTL*	*CCY*	*CT*	*PREREQ*
CSC222	*C++ Programming*	*50*	*3*	*CSC221*
CSC542	*Database Design*	*25*	*3*	*CSC222*
CSC548	*Object Oriented Design*	*25*	*3*	*CSC542*
CSC308	*Visual Basic*	*40*	*3*	*CSC221*

Note that CRS_NBR is the primary key of the relation CRS. The DOM(CRS_NBR) is the set of all course numbers. Now, DOM(PREREQ) is the set of all course numbers. Thus PREREQ is a foreign key of the relation CRS referencing the relation CRS.

Note that course number CSC108 appears as prerequisite. However, there is no course listed with CSC108 as course number. This does not negate the assertion that PREREQ is a foreign key of CRS. Rather, this is a data integrity issue that is addressed next.

The second data integrity rule, also known as, **referential integrity rule** can be stated as follows.

A foreign key entry can be null or an entry that matches a primary key value of the referencing table.

The purpose of this integrity rule is to ensure that every entity referred in fact exists in the referencing table. We allow null values, as a practical consideration. For example, consider the Example 2.2.9. A prospective customer may call the insurance agency. At this point, he or she has completely decided which agent to choose. Therefore, it is better to leave the AGNT attribute null for the time being. In order to allow similar flexibility, a foreign key is permitted to have a null value, provided it does not violate entity integrity rule.

Every relation in a database must obey the referential integrity rule.

Example 2.2.11 *Continuing with Example 2.2.9, recall that AGNT is a foreign key and it references AGENT relation. The set of non-null values of AGNT in relation CUSTOMER is {1,2}. The set of values of the attribute AGNT_NBR, the primary key of the referencing table, is {1,2,3,4}. Clearly, {1,2} is a subset of {1,2,3,4}. Therefore, as far as the foreign key AGNT is concerned, the database obeys the referential integrity rule.*

Example 2.2.12 *In this example, we continue with Example 2.2.10. We have observed that PREREQ is a foreign key referencing CRS. The set of values of the attribute CRS_NBR, the primary key of the referencing table, is {CSC221, CSC222, CSC542, CSC548, CSC 308}. The set of non-null values of PREREQ in relation CRS is {CSC221,CSC222, CSC542, CSC108} and is not a subset of {CSC221,CSC222, CSC542, CSC548, CSC 308}. Therefore, database violates the referential integrity rule.*

A secondary key is defined as a key that is used strictly for data retrieval purposes. Consider the CUSTOMER table presented in Table 2.4. A customer may not remember the customer identification number, as he or she makes her phone call. Data retrieval for a customer can be achieved when customer's last name or phone number is used. In this case, the primary key is customer identification number. The secondary key is the pair of attributes customer's last name and phone number. The secondary key, in general, need not be a unique identifier. However, it narrows down the search to a more manageable number.

A secondary key's usefulness depends on many factors. First, it must be restrictive enough to narrow down the search. Secondly, the attributes should be easily remembered by the user. For example, consider the CUSTOMER table given in Table 2.4. A customer remembers his or her last name and phone number all the time. Other attributes, may lead to confusion. For example, as address such as *8807 N 136th street* can be entered as *8807 North 136th street* or *8807 N 136 St.* and so on.

A data dictionary is used to keep a detailed accounting of all tables within the database. Thus the data dictionary contains information about all attributes and their characteristics. It may also contain information about whether or not an attribute is a primary key, whether or not an attribute must have a non-null value. It may also contain information about foreign keys.

The system catalog can be thought of as a database on all objects in the database. Thus for example, system catalog contains information about authorized users, their privileges and so on. In fact all current implementations of relational database management systems contain not only the system catalog information but also the data dictionary information can be derived from the system catalog. The system catalog provides all necessary documentation in a natural way.

As a new table is created, system catalog helps the RDBMS to check for and eliminate homonyms and synonyms. Homonyms are similar sounding words with different meanings, such as peace and piece; or identically spelled words with different meanings, such as right (as correct) and right (as in right turn). In database context, we may use LNAME to denote customer last name in CUSTOMER table and LNAME to denote the employee last name in EMPLOYEE table. To avoid confusion, one should try to use homonyms. Synonyms are different words with the same meaning. For example, in context of certain database two different words car and auto may refer to the same object. Using car as an attribute in one table and auto as an attribute in another table to refer to the same object may lead to confusion. In other words, a foreign key attribute should share the same name with the primary key of the referencing table.

2.3 The Relational Operators

In our every day life, we deal with real numbers. We have four basic operators, $+, -, \times$ and $\div$. These operators possess certain important properties. In particular, the following properties deserve special mention in this context.

Closure property: If we add two real numbers, the result we get is another real number; not a character or a string or a date. This is known as the closure property of addition. It is well known that $+, -$ and $\times$ also enjoys closure property. In the case of division operation, division by zero is not allowed. In all other cases, division operation has the closure property.

Algebraic relations: Addition and multiplication are commutative and hence, $a+b = b+a$ and $a \times b = b \times a$ for all real numbers a and b. Addition and multiplication are associative. Thus, $a + (b + c) = (a + b) + c$ and $a \times (b \times c) = (a \times b) \times c$ for all real numbers a, b and c. Further, multiplication is distributive over addition. Hence, $a \times (b + c) = a \times b + a \times c$ for all real numbers a, b and c.

The above results present us with two levels of opportunities for optimizing a computation. First, we can rearrange the terms so that the number of operations performed can be made less. For example, if we want to compute $a \times b + b \times c + c \times d$ and we proceed to compute as such, we have to perform three multiplications and two additions. Instead, we can rewrite the above expression as $a \times b + c \times (b + d)$. Now, we need to perform two multiplications and two additions. Thus there is one less multiplication operation!

Another level of optimization is possible at the operation implementation level. For example, if one could speed up the multiplication operation, resulting cost savings will affect all numeric computations that involve at least one multiplication. There are only four basic operators. Therefore, it is reasonable to spend resources to speed up these four basic operations and thus try to improve the performance of the system as a whole.

In this section, we introduce a small set of relational operators and try to imitate the real number system. In other words, we would like to have a small set of relational operators such that with closure properties. This will ensure that relational operators can be applied one after another as in the case of real numbers. We would also like to have some commutative laws, some associative laws and some distributive laws, so that we can rewrite a relational expression to an equivalent less costly expression.

Union, intersection and difference

Two relational schemes are said to be *union compatible* if they have the same attribute characteristics. More formally, two relational schemes $R = \{A_1, A_2, \ldots, A_n\}$ and $S = \{B_1, B_2, \ldots, B_m\}$ are union compatible if the following conditions hold:

1. $n = m$ or the number of attributes is the same;
2. $DOM(A_i) = DOM(B_i)$, for $i = 1, 2, \ldots, n$.

Let REL_R and REL_S be two relations on relational schemes R and S respectively. Relational operations, union, intersection and difference on REL_R and REL_S are defined, if and only if R and S are union compatible.

We use the following two union compatible relations to illustrate these operations.

REL_R

A	B	C
a_1	b_1	c_1
a_1	b_1	c_2
a_2	b_2	c_1
a_2	b_1	c_2

REL_S

A	B	C
a_1	b_2	c_2
a_2	b_2	c_1
a_1	b_1	c_2

Union

The *union* is a binary operation. When applied to two union compatible relations, it yields a new relation that is union compatible to both of them. The union of REL_R and REL_S is obtained by combining the rows of both REL_R and REL_S. We use the notation $REL\_R \cup REL\_S$ to denote the union of REL_R and REL_S.

$REL\_R \cup REL\_S$

A	B	C
a_1	b_1	c_1
a_1	b_1	c_2
a_2	b_2	c_1
a_2	b_1	c_2
a_1	b_2	c_2

The tuple $(a1, b1, c2)$ appears in both relations REL_R and REL_S. However, it appears only once in $REL\_R \cup REL\_S$.

Intersection

The intersection is a binary operation. When applied to two union compatible relations, it yields a new relation that is union compatible to both of them. The intersection of REL_R and REL_S is obtained by taking rows common to both REL_R and REL_S. The notation $REL\_R \cap REL\_S$ is used to denote the intersection of REL_R and REL_S.

$REL\_R \cap REL\_S$

A	B	C
a_1	b_1	c_2
a_2	b_2	c_1

Difference

The *difference* is a binary operation. When applied to two union compatible relations, it yields a new relation that is union compatible to both of them. We use the notation $REL\_R - REL\_S$ to denote the difference of REL_R from REL_S. The result is obtained by removing all the tuples common to R and S from R. Therefore, $REL\_R - REL\_S = REL\_R - (REL\_R \cap REL\_S)$.

$REL\_R - REL\_S$

A	B	C
a_1	b_1	c_1
a_2	b_1	c_2

Select

The *select* is a unary operation. When applied to a relation, it yields a new union compatible relation. The select operator produces a new relation by selecting tuples that satisfy a given predicate. We use the symbol σ to denote the select operation.

To be more specific, let REL_R be a relation on a relational scheme $R = \{A, B, C\}$. For $a \in DOM(A), \sigma_{A\,=\,a} REL\_R$ is a select operation on REL_R that will create a new relation having all tuples of REL_R with attribute A equal to a.

$\sigma_{A\,=\,a_1} REL\_R$

A	B	C
a_1	b_1	c_1
a_1	b_1	c_2

Further, if we would like to select all tuples of REL_R with attribute A equal to a_1 and attribute B equal to b_2 where $a_1 \in DOM(A), b_2 \in DOM(B)$, can be computed by the expression $\sigma_{(A\,=\,a_1) \cup (B\,=\,b_2)} REL\_R$.

$\sigma_{(A\,=\,a_1) \cup (B\,=\,b_2)} REL\_R$.

A	B	C
a_1	b_1	c_1
a_1	b_1	c_2
a_2	b_2	c_1

Example 2.3.1 *Consider the CRS relation given below.*

CRS

CRS_NBR	CRS_TTL	CCY	CT	PREREQ
CSC221	C Programming	50	3	
CSC222	C++ Programming	50	3	CSC221
CSC542	Database Design	25	3	CSC222
CSC548	Object Oriented Design	25	3	CSC542
CSC308	Visual Basic	40	1	CSC221

All courses with CSC221 as prerequisites can be found as follows:

$\sigma_{PREREQ=CSC221}CRS$

CRS_NBR	CRS_TTL	CCY	CT	PREREQ
CSC222	C++ Programming	50	3	CSC221
CSC308	Visual Basic	40	1	CSC221

Courses with CSC221 as prerequisite and capacity less than 50 can found using the select operator as shown below.

$\sigma_{(PREREQ\ =\ CSC221)\cap(CCY\ <\ 50)}CRS$

CRS_NBR	CRS_TTL	CCY	CT	PREREQ
CSC308	Visual Basic	40	1	CSC221

Courses with CSC221 or CSC222 as a prerequisite are obtained as follows.

$\sigma_{(PREREQ\ =\ CSC221)\cap(PREREQ\ =\ CSC222)}CRS$

CRS_NBR	CRS_TTL	CCY	CT	PREREQ
CSC222	C++ Programming	50	3	CSC221
CSC542	Database Design	25	3	CSC222
CSC308	Visual Basic	40	1	CSC221

Project

The project is a unary operation. The project operation discards unwanted attributes. We use the symbol Π to denote the project operation.

Let REL_R be a relation on a relational scheme $R = \{A, B, C\}$. For $X \subseteq \{A, B, C\}, \Pi_X REL\_A$ is defined as a relation on relational scheme X such that t is a tuple in REL_R if and only if $t(X)$ is a tuple in $\Pi_X REL\_A$. For example, if $X = \{A, C\}$, we have the following:

$\Pi_{\{A,C\}} REL\_A$

A	C
a_1	c_1
a_1	c_2
a_2	c_1
a_2	c_2

Example 2.3.2 *Consider the CRS relation in Example 2.3.1. If we need the course number, course title and prerequisite, the project operator can be used as follows:*

$\Pi_{\{CRS\_NBR, CRS\_TTL, PREREQ\}} CRS$

CRS_NBR	*CRS_TTL*	*PREREQ*
CSC221	*C Programming*	
CSC222	*C++ Programming*	*CSC221*
CSC542	*Database Design*	*CSC222*
CSC548	*Object Oriented Design*	*CSC542*
CSC308	*Visual Basic*	*CSC221*

Example 2.3.3 *In this example, we combine both select and project operators. Assume that we need to find the course number and course title of all the courses with prerequisite CSC221. This can be accomplished by* $\Pi_{\{CRS\_NBR, CRS\_TTL\}}(\sigma_{PREREQ=CSC221} CRS)$.
First, $\sigma_{PREREQ=CSC221} CRS$ *selects all tuples with* $PREREQ = CSC221$. *Then, the project operator eliminates all attributes other than* CRS_NBR *and* CRS_TTL. *Thus we have the following.*

$\Pi_{\{CRS\_NBR, CRS\_TTL\}}(\sigma_{PREREQ=CSC221} CRS)$

CRS_NBR	*CRS_TTL*
CSC221	*C Programming*
CSC222	*C++ Programming*
CSC542	*Database Design*
CSC548	*Object Oriented Design*
CSC308	*Visual Basic*

Product

The *product* is a binary operation. Let REL_R and REL_T be relations on relational schemes $R = \{A_1, A_2, \ldots, A_n\}$ and $T = \{B_1, B_2, \ldots, B_m\}$ respectively. Then the product of REL_R and REL_T is a relation on $\{A_1, A_2, \ldots, A_n, B_1, B_2, \ldots, B_m\}$. The product produces a list of all possible pairs of tuples from two relations. Therefore, if one relation has five rows and the other has 10 rows, the relation obtained by taking their product has 50 tuples. We use the notation $\times$ for the product operator.

Let REL_R be a relation on $R = \{A, B, C\}$ and REL_T be a relation on $T = \{D, E\}$.

REL_R

A	B	C
a_1	b_1	c_1
a_1	b_1	c_2
a_2	b_2	c_1
a_2	b_1	c_2

REL_T

D	E
d_1	e_2
d_2	e_2
d_1	e_1

Now $REL\_R \times REL\_T$ is a relation on $\{A, B, C, D, E\}$ and has 12 tuples. It may be noted that each tuple is produced by one tuple from REL_R and REL_T.

$REL\_R \times REL\_T$

A	B	C	D	E
a_1	b_1	c_1	d_1	e_2
a_1	b_1	c_2	d_1	e_2
a_2	b_2	c_1	d_1	e_2
a_2	b_1	c_2	d_1	e_2
a_1	b_1	c_1	d_2	e_2
a_1	b_1	c_2	d_2	e_2
a_2	b_2	c_1	d_2	e_2
a_2	b_1	c_2	d_2	e_2
a_1	b_1	c_1	d_1	e_1
a_1	b_1	c_2	d_1	e_1
a_2	b_2	c_1	d_1	e_1
a_2	b_1	c_2	d_1	e_1

Join

The *join* or *natural join* is a binary operation. Let REL_R and REL_T be two relations on relational schemes

$$R = \{A_1, A_2, \ldots, A_n, C_1, C_2, \ldots, C_k\}$$

and

$$T = \{B_1, B_2, \ldots, Bm, C_1, C_2, \ldots, C_k\}$$

respectively such that $\{C1, C2, \ldots, Ck\}$ is a set of attributes common to R and T. Join links tables REL_R and REL_T by selecting only rows with identical values in their common attribute(s) and produces a relation on

$$\{A1, A2, \ldots, An, C1, C2, \ldots, Ck, B1, B2, \ldots, Bm\}.$$

We use the symbol $\bowtie$ for the natural join operator. In the special case where R and T have no common attributes,

$$REL\_R \bowtie REL\_T = REL\_R \times REL\_T.$$

Informally, natural join computation can be thought of as a three-step process. First we compute the product. The second step involves selecting rows with identical values in their common attribute(s). Since common attributes are repeated, we delete repeating attributes in the third step.

Let REL_R be a relation on $R = \{A, B, C\}$ and REL_T be a relation on $T = \{D, C, E\}$.

REL_R

A	B	C
a_1	b_1	c_1
a_2	b_1	c_2
a_3	b_2	c_1
a_4	b_1	c_3

REL_T

D	C	E
d_1	c_1	e_2
d_2		e_3
d_3	c_2	e_2
d_4	c_2	e_1

Step 1: Compute the product

In this case, there is only one common attribute, C. In what follows, we use $REL\_R.C$ to denote the attribute C in R and use $REL\_T.C$ to denote the attribute C in T respectively. In this step, we compute the product of REL_R and REL_T.

$REL\_R \times REL\_T$

A	B	$REL\_R.C$	D	$REL\_T.C$	E
a_1	b_1	c_1	d_1	c_1	e_2
a_2	b_1	c_2	d_1	c_1	e_2
a_3	b_2	c_1	d_1	c_1	e_2
a_4	b_1	c_3	d_1	c_1	e_2
a_1	b_1	c_1	d_2		e_3
a_2	b_1	c_2	d_2		e_3
a_3	b_2	c_1	d_2		e_3
a_4	b_1	c_3	d_2		e_3
a_1	b_1	c_1	d_3	c_2	e_2
a_2	b_1	c_2	d_3	c_2	e_2
a_3	b_2	c_1	d_3	c_2	e_2
a_4	b_1	c_3	d_3	c_2	e_2
a_1	b_1	c_1	d_4	c_2	e_1
a_2	b_1	c_2	d_4	c_2	e_1
a_3	b_2	c_1	d_4	c_2	e_1
a_4	b_1	c_3	d_4	c_2	e_1

Step 2: Select rows with identical values in their common attribute(s)

$\sigma_{REL\_R.C=REL\_T.C} REL\_R \times REL\_T$

A	B	$REL\_R.C$	D	$REL\_T.C$	E
a_1	b_1	c_1	d_1	c_1	e_2
a_3	b_2	c_1	d_1	c_1	e_2
a_2	b_1	c_2	d_3	c_2	e_2
a_2	b_1	c_2	d_4	c_2	e_1

Step 3: Project on distinct attribute(s)

Attributes $REL\_R.C$ and $REL\_T.C$ are identical. Therefore, we eliminate the repeating attribute, say $REL\_T.C$. Further, we rename $REL\_R.C$ as C. Thus

$REL\_R \bowtie REL\_T$

A	B	C	D	E
a_1	b_1	c_1	d_1	e_2
a_3	b_2	c_1	d_1	e_2
a_2	b_1	c_2	d_3	e_2
a_2	b_1	c_2	d_4	e_1

Example 2.3.4 *Consider the relations AGENT and CUSTOMER maintained by an insurance agency.*

AGENT

A_FNAME	A_LNAME	ID	AGNT PHONE
Molly	Smith	1	4444568
Mike	Covey	2	5785678
Sue	Newman	3	6489490
Chris	Levy	4	3451919

CUSTOMER

FNAME	LNAME	C_NBR	AMT	AGNT
John	Shaw	56326	100,000	2
Meera	Nair	67674	2,000,000	3
Magan	Fields	56371	400,000	1
⋮	⋮	⋮	⋮	⋮
Joyce	Lendl	83429	750,000	2

Let us assume that we want to create a table with the customer name followed by the agent name. Therefore, we have to first combine CUSTOMER table with AGENT table. This is accomplished by joining both relations. This table being too wide, we display it by breaking it in to two. It must be noted that $CUSTOMER \bowtie AGENT$ relation has eight attributes. We display four of them first, followed by the remaining four.

$CUSTOMER \bowtie AGENT$

FNAME	LNAME	C_NBR	AMT
John	Shaw	56326	100,000
Meera	Nair	67674	2,000,000
Magan	Fields	56371	400,000
⋮	⋮	⋮	⋮
Joyce	Lendl	83429	750,000

AGNT	A_FNAME	A_LNAME	AGNT PHONE
2	Mike	Covey	5785678
3	Sue	Newman	6489490
1	Molly	Smith	4444568
⋮	⋮	⋮	⋮
2	Mike	Covey	5785678

$\Pi_{\{FNAME,LNAME,A\_FNAME,A\_LNAME\}} CUSTOMER \bowtie AGENT$ *produces the desired result.*

FNAME	*LNAME*	*A_FNAME*	*A_LNAME*
John	*Shaw*	*Mike*	*Covey*
Meera	*Nair*	*Sue*	*Newman*
Magan	*Fields*	*Molly*	*Smith*
⋮	⋮	⋮	⋮
Joyce	*Lendl*	*Mike*	*Covey*

Equi-join and theta-join

Both *equi-join* and *theta-join* are binary operations. Let REL_R, REL_T be two relations on relational schemes $R = \{A_1, A_2, \ldots, A_n\}$ and $T = \{B_1, B_2, \ldots, B_m\}$ respectively. Equi-join links tables REL_R and REL_T on the basis of an equality condition that compares specified attributes and produces a relation on $\{A_1, A_2, \ldots, A_n, B_1, B_2, \ldots, B_m\}$. The attributes involved in comparison must have identical domains. We use the symbol $[attribute_1 = attribute_2,]$ for the equi-join operator. Unlike join operator, duplicate attributes are kept in this case. If the comparison operator is anything other than equality, it is called a theta-join.

Informally, equi-join (theta-join) computation can be thought of as a two step process. First we compute the product. The second step involves selecting rows satisfying the condition specified. We now proceed to illustrate the equi-join operator.

Let REL_R be a relation on $R = \{A, B, C1\}$ and REL_T be a relation on $T = \{D, C2, E\}$, where $DOM(C1) = DOM(C2)$.

REL_R

A	B	$C1$
a_1	b_1	c_1
a_2	b_1	c_2
a_3	b_2	c_1
a_4	b_1	c_3

REL_T

D	$C2$	E
d_1	c_1	e_2
d_2		e_3
d_3	c_2	e_2
d_4	c_2	e_1

Step 1: Compute the product
$REL\_R \times REL\_T$

A	B	$C1$	D	$C2$	E
a_1	b_1	c_1	d_1	c_1	e_2
a_2	b_1	c_2	d_1	c_1	e_2
a_3	b_2	c_1	d_1	c_1	e_2
a_4	b_1	c_3	d_1	c_1	e_2
a_1	b_1	c_1	d_2		e_3
a_2	b_1	c_2	d_2		e_3
a_3	b_2	c_1	d_2		e_3
a_4	b_1	c_3	d_2		e_3
a_1	b_1	c_1	d_3	c_2	e_2
a_2	b_1	c_2	d_3	c_2	e_2
a_3	b_2	c_1	d_3	c_2	e_2
a_4	b_1	c_3	d_3	c_2	e_2
a_1	b_1	c_1	d_4	c_2	e_1
a_2	b_1	c_2	d_4	c_2	e_1
a_3	b_2	c_1	d_4	c_2	e_1
a_4	b_1	c_3	d_4	c_2	e_1

Step 2: Select rows with identical values in the indicated attribute(s)
$REL\_R[C1 = C2]REL\_T$

A	B	$C1$	D	$C2$	E
a_1	b_1	c_1	d_1	c_1	e_2
a_3	b_2	c_1	d_1	c_1	e_2
a_2	b_1	c_2	d_3	c_2	e_2
a_2	b_1	c_2	d_4	c_2	e_1

Divide

Divide operation is very similar to integer division operation. Consider the integer division of 17 by 5. The result is 3. There are two important points worth considering in this regard.

- $17 \geq 5 \times 3$;
- $17 < 5 \times n$, for all integers $n > 3$.

Thus 3 is the largest integer value of n such that $17 \geq 5 \times n$.

Divide is a binary operator and we will use the notation $\div$. Let REL_R and REL_T be relations on relational schemes
$R = \{A1, A2, \ldots, An, C1, C2, \ldots, Ck\}$
and
$T = \{C1, C2, \ldots, Ck\}$

respectively. Then $REL\_S = REL\_R \div REL\_T$ is defined and is a relation on the relational scheme $S = \{A1, A2, \ldots, An\}$ such that the following conditions hold.

- $REL\_R \supseteq REL\_T \times REL\_S$;
- $REL\_R \subset REL\_T \times REL\_Q$, for all relations $REL\_Q \supset REL\_S$.

Informally, a divide computation can be thought of as a three-step process. First for each tuple t of REL_T, select all tuples r of REL_R such $r(T) = t$. Recall that T is the relational scheme of REL_T. Let us denote the relation so obtained by REL_R_t. The second step involves projecting all relations REL_R_t on attributes of S. Thus by the end of step 2, for each tuple t of REL_T, we have a relation $\Pi_S REL\_R_t$ on the relational scheme S. The final step is the intersection of all relations $\Pi_S REL\_R_t$.

Let REL_R be a relation on $R = \{A, B\}$ and REL_T be a relation on $T = \{A\}$.

REL_R

A	B
a_1	b_1
a_2	b_1
a_3	b_2
a_4	b_3
a_4	b_8
a_5	b_8
a_2	b_2
a_1	b_4
a_1	b_2

REL_T

A
a_1
a_2

Step 1: For each tuple t of REL_T, select all tuples r of REL_R such $r(T) = t$.

Let $t = a_1$. Select tuples of REL_R such that attribute A is a_1. That is, $\sigma_{A=a_1} REL\_R$.

$\sigma_{A=a_1} REL\_R$

A	B
a_1	b_1
a_1	b_4
a_1	b_2

Now choose t as a_2. Select tuples of REL_R such that attribute A is a_2. That is, $\sigma_{A=a_2} REL\_R$.

$\sigma_{A=a_2} REL\_R$

A	B
a_2	b_1
a_2	b_2

Step 2: Project all relations $\sigma_{A=a_i} REL\_T$ on attributes of S.
In this case, $S = \{B\}$, the set of attributes that are in REL_R but not in REL_T. Thus we have the following.

$\Pi_B \sigma_{A=a_1} REL\_T$

B
b_1
b_4
b_2

and

$\Pi_B \sigma_{A=a_2} REL\_R$

B
b_1
b_2

Step 3: Compute the intersection of all relations of the form $\Pi_B \sigma_{A=a_i} REL\_R$.
It is clear that intersection of $\Pi_B \sigma_{A=a_1} REL\_R$ and $\Pi_B \sigma_{A=a_2} REL\_R$ is the following:

$REL\_R \div REL\_T$

B
b_1
b_2

Example 2.3.5 *The following example may illustrate the use of the division operation. Here is a list of people and the states they have visited in the past.*

VISITED

PERSON	*STATE*
John Ott	*California*
John Ott	*Iowa*
John Ott	*New York*
Chris Lee	*California*
Chris Lee	*Nebraska*
Peter Witt	*New York*
Peter Witt	*California*
Mark Mayer	*Iowa*
Mark Mayer	*Florida*

If we need to find all states visited by a set of people, say, John Ott and Peter Witt. One way is first form a relation involving the set of people PRSN as follows.

PRSN

PERSON
John Ott
Peter Witt

Now, we divide VISITED by PRSN to obtain the following.

VISITED $\div$ *PRSN*

STATE
California
New York

It is worth noticing that California and New York are the states visited by both John Ott and Peter Witt.

3

THE CISET RELATIONAL MODEL

In this chapter we introduce the ciset relational model, a collection of methods and techniques for organizing data centered on the notion of a ciset relation as the data structure. After we present the notion of a ciset relation, we present other fundamental concepts such as candidate key, primary key and foreign key in this model. Both entity integrity rule and the referential integrity rule are also explored.

A ciset relational model is a blueprint of the database that can store conflicting information. In a ciset relational model, the data always is presented in the form of a table, which we call a *ciset relation*. Every piece of information is stored in one or more tables. Informally, a ciset relational database can be conceived as a collection of ciset relations. For example, data on all professors of a university can be organized in the form of a table FACULTY as shown in Table 3.1.

TABLE 3.1 ciset relation: FACULTY

FACULTY

F_ID	F_NAME	DEPT	EVALUATION
12312	John Smith	Marketing	$\langle 0.5, 0.7 \rangle$
31897	Mary Lee	Mathematics	$\langle 0.4, 0.9 \rangle$
12674	Sandy Dewitt	Marketing	$\langle 0.1, 0.8 \rangle$
56739	Bea Anthony	Accounting	$\langle 0.7, 0.6 \rangle$

As we contemplate the FACULTY table, we can identify several elements.

- The name of the table: FACULTY.
- The column headings of the table: F_ID, F_NAME,DEPT and EVALUATION.
- The set of rows of the table.
- The column headings of table are known as *ciset relational attributes.* Thus the FACULTY table has four ciset relational attributes: F_ID, F_NAME, DEPT and EVALUATION.
- Each row of the table is known as a *ciset tuple*. Thus the above table has four ciset tuples.

From the user perspective, database model is still the classical relational model. The only conceptual difference as far as the user is concerned can be summarized as follows.

1. As user inserts a new piece of data, user is asked to provide the confidence level. Similar is the situation when the user tries to delete a piece of information or modify a piece of information.
2. In addition to producing results of a query user can use the confidence index attribute to obtain the level of "trust" one can place on the result itself.

3.1 Formalization of Ciset Relation

We now proceed to formalize the notion of a table. Let U be the set of all ciset relational attributes. For each attribute $A \in U$, let $DOM(A)$, called domain of A, denote the set of all possible values that can occur in that column. The domains are arbitrary, nonempty sets, nonempty fuzzy sets or nonempty cisets or nonempty subset of confidence indexes, finite or countably infinite. Note that if there is only one element in $DOM(A)$, we need not include A as an attribute of the table. It is safe to assume that $DOM(A)$ contains at least two different values.

Let $R = \{A_1, \ldots, A_n\}$ be a finite set of *ciset relational attributes.* Then R is called a *ciset relational scheme.* A *ciset relation* r on a ciset relational scheme R is a finite set of mappings $\{t_1, \ldots, t_m\}$ from R to $\cup\{DOM(A_i) \mid i = 1, \ldots, n.\}$ with the restriction that $t(A_i) \in DOM(A_i), i = 1, \ldots, n$ for all $t \in \{t_1, \ldots, t_m\}$. The mappings are called *ciset tuples.*

Depending on the complexity of $DOM(A_i)$, $i = 1, \ldots, n$, ciset relations can be classified into many categories, namely, Type 0, Type 1, Type 2, Type 3 and so on.

- In Type 0 ciset relation, $DOM(A_i)$, $i = 1, \ldots, n$, is a set. Thus in addition to the commonly used data types such as numbers, strings, Boolean values and dates, a Type 0 ciset relation can accommodate confidence index. Therefore, $DOM(A_i)$ can be a subset of $\mathfrak{C}$.

- In Type 1 ciset relation, $DOM(A_i)$, $i = 1, \ldots, n$, is a ciset. Note that ciset is a generalization of sets and fuzzy sets. Thus in the case of a Type 1 ciset relation, $DOM(A_i)$, $i = 1, \ldots, n$, is a set, or a fuzzy set or a ciset.

- In Type 2 ciset relation, $DOM(A_i)$, $i = 1, \ldots, n$, is a set of subsets of a ciset. In the case of a Type 2 ciset relation, $DOM(A_i)$, $i = 1, \ldots, n$, is a set of subsets of a set, or a set of subsets a fuzzy set or a set of subsets a ciset.

- In Type 3 ciset relation, $DOM(A_i)$, $i = 1, \ldots, n$, is a ciset of subsets of a ciset. Thus in the case of a Type 3 ciset relation, $DOM(A_i)$, $i = 1, \ldots, n$, is a set or a fuzzy set or a ciset of subsets of sets; or set or a fuzzy set or a ciset of subsets fuzzy sets; or a set or a fuzzy set or a ciset of subsets of cisets.

- In Type $2j$ ciset relation, $DOM(A_i)$, $i = 1, \ldots, n$, is a set of subsets of Type $(2j - 1)$ domain, $j > 1$.

- In Type $2j + 1$ ciset relation, $DOM(A_i)$, $i = 1, \ldots, n$, is a ciset of subsets of Type $(2j - 1)$ domain, $j > 1$.

The classification can be summarized as shown in Table 3.2.

TABLE 3.2 Classification of ciset relations

Type	$DOM(A_i)$
0	set
1	ciset
2	set of subsets of a ciset
3	ciset of subsets of a ciset
⋮	⋮
2j	set of subsets of Type $(2j - 1)$ domain, $j > 1$.
2j + 1	ciset of subsets of Type $(2j - 1)$ domain, $j > 1$

Example 3.1.1 *In Table 3.1, the ciset relational scheme is* $FCLY\_SCHEME = \{F\_ID, F\_NAME, DEPT, EVALUATION\}$. *The tuple* $t = (12312, John\ Smith, Marketing, \langle 0.5, 0.7 \rangle)$ *can be considered as a mapping defined on the set* $\{F\_ID, F\_NAME, DEPT, EVALUATION\}$ *by*

$t(F\_ID) = 12312$,
$t(F\_NAME) = John\ Smith$,
$t(DEPT) = Marketing$,
$t(EVALUATION) = \langle 0.5, 0.7 \rangle$.

The FACULTY ciset relation has four ciset tuples and four ciset attributes. The domain for each of these attributes is as follows.

$DOM(F\_ID)$ = *set of all five digit numbers,*
$DOM(F\_NAME)$ = *set of all names,*
$DOM(DEPT)$ = *{Marketing, Mathematics, Accounting},*
$DOM(EVALUATION) = \mathfrak{C}$.

Thus $DOM(F\_ID), DOM(F\_NAME), DOM(DEPT)$ *and* $DOM(EVALUATION)$ *satisfies the domain requirements of a Type 0 ciset relation. Therefore we classify FACULTY as a Type 0 ciset relation.*

From a practical point of view, it is quite easy to implement a Type 0 ciset relation as opposed to Type 1, Type 2 or Type i (i > 2) ciset relations. At this point there is no commercially available fuzzy database. Thus to hope for a commercially available ciset database may be too much to wish for. On the other hand, it is possible to simulate a Type 0 ciset relation in the current technology. So, we concentrate on Type 0 ciset relational database for the rest of this book.

Semantics of a ciset relation

Consider the following Type 0 ciset relation given in Table 3.3.

TABLE 3.3 ciset relation: FACULTY

FACULTY

F_ID	F_NAME	DEPT	EVALUATION	CI
12312	John Smith	Marketing	$\langle 0.5, 0.7 \rangle$	$\langle 0, 1 \rangle$
31897	Mary Lee	Mathematics	$\langle 0.4, 0.9 \rangle$	$\langle 0.2, 0.9 \rangle$
12674	Sandy Dewitt	Marketing	$\langle 0.1, 0.8 \rangle$	$\langle 0.3, 0.7 \rangle$
56739	Bea Anthony	Accounting	$\langle 0.7, 0.6 \rangle$	$\langle 0.2, 0.9 \rangle$

The semantics of the first tuple is "John Smith is a faculty member of Marketing department and his faculty identification number is 12312. John Smith's student evaluation is $\langle 0.5, 0.7 \rangle$. The fact that John Smith is a faculty member of Marketing department and his faculty identification number is 12312 has a confidence index value $\langle 0, 1 \rangle$." Note that there is a subtle difference between the ways we interpreted attributes EVALUATION and CI. The attribute EVALUATION is in all respects, just like any other attribute and asserts a fact about John Smith. On the other hand, the attribute CI is not asserting a fact about John Smith. *Instead, CI asserts a fact about the validity of the tuple itself and assigns a confidence index for the tuple.* Thus CI plays a very different role compared to all other attributes. The attribute CI is a *tuple attribute that qualifies the tuple* and all other attributes are *entity attributes.*

Let $R = \{A_1, \ldots, A_n\}$ be a ciset relational scheme. Then there can be at most one attribute that measures the validity of the tuple itself. This being a special attribute, for the rest of this book, we shall refer it as CI and we call it a *tuple attribute.* The set $\mathcal{E}(R) = \{A_i \mid A_i \neq CI\}$ is called the set of *entity attributes* of R and the set $\mathcal{R}(R) = \{A_i \mid DOM(A_i) \cap \mathfrak{C} = \varnothing\}$ is called the set of *relational attributes* of R.

Given a ciset relation r on a ciset relational scheme R, let $\mathcal{E}(r)$ and $\mathcal{R}(r)$ denote ciset relations on $\mathcal{E}(R)$ and $\mathcal{R}(R)$ respectively. Note that $\mathcal{R}(r)$ degenerates to a traditional relation since all attributes of $\mathcal{R}(R)$ are relational attributes. Let t be a tuple of r. Then we use the notation $t(\mathcal{E})$ and $t(\mathcal{R})$ to denote the corresponding tuples in $\mathcal{E}(R)$ and $\mathcal{R}(R)$ respectively. Further, by abusing the notation, we say $t = (t(\mathcal{E}), t(CI))$.

In Table 3.3, the ciset relational scheme is $FACULTY\_SCHEME = \{F\_ID, F\_NAME, DEPT, EVALUATION, CI\}$. The tuple attribute of $FACULTY\_SCHEME$ is CI. The entity attributes of
$FACULTY\_SCHEME$ is $\mathcal{E}(FACULTY\_SCHEME)$
$= \{F\_ID, F\_NAME, DEPT, EVALUATION\}$
and the set of relational attributes of $FACULTY\_SCHEME$ is
$\mathcal{R}(FACULTY\_SCHEME) = \{F\_ID, F\_NAME, DEPT\}$.
Further,

$\mathcal{E}$(FACULTY)

F_ID	F_NAME	DEPT	EVALUATION
12312	John Smith	Marketing	$\langle 0.5, 0.7\rangle$
31897	Mary Lee	Mathematics	$\langle 0.4, 0.9\rangle$
12674	Sandy Dewitt	Marketing	$\langle 0.1, 0.8\rangle$
56739	Bea Anthony	Accounting	$\langle 0.7, 0.6\rangle$

and

$\mathcal{R}$(FACULTY)

F_ID	F_NAME	DEPT
12312	John Smith	Marketing
31897	Mary Lee	Mathematics
12674	Sandy Dewitt	Marketing
56739	Bea Anthony	Accounting

.

Let $t = (12312, John\ Smith, Marketing, \langle 0.5, 0.7\rangle, \langle 0, 1\rangle)$. Then
$t(\mathcal{E}) = (12312, John\ Smith, Marketing, \langle 0.5, 0.7\rangle)$
and
$t(\mathcal{R}) = (12312, John\ Smith, Marketing)$.
Further,
$t = (12312, John\ Smith, Marketing, \langle 0.5, 0.7\rangle, \langle 0, 1\rangle) = (t(\mathcal{E}), t(CI))$ where $t(CI) = \langle 0, 1\rangle$.

As noted earlier, the tuple
$t = (12312, John\ Smith, Marketing, \langle 0.5, 0.7\rangle, \langle 0, 1\rangle)$
can be considered as a mapping defined on the set $FACULTY\_SCHEME$ by

$t(F\_ID) = 12312,$
$t(F\_NAME) = John\ Smith,$
$t(DEPT) = Marketing,$
$t(EVALUATION) = \langle 0.5, 0.7\rangle,$
$t(CI) = \langle 0, 1\rangle.$

On the other hand the tuple
$t = (12312, John\ Smith, Marketing, \langle 0.5, 0.7\rangle, \langle 0, 1\rangle)$
can also be considered as a ciset on $\mathcal{E}(FACULTY)$ or equivalently, a mapping from $\mathcal{E}(FACULTY)$ in to $\mathfrak{C}$. For the rest of this book, we shall make use of this fact.

Properties of a ciset relation

Property 1: The order of attributes is immaterial

One of the important consequences of considering tuples as mappings is that the order of attributes is immaterial, from a theoretical point of view. We now proceed to show that the order of attributes is immaterial, from a practical point of view. For example, the following two ciset relations are the same.

Consider two ciset relations FACULTY and FCLTY given below.

FACULTY

F_ID	F_NAME	DEPT	EVALUATION	CI
12312	John Smith	Marketing	$\langle 0.5, 0.7\rangle$	$\langle 0, 1\rangle$
31897	Mary Lee	Mathematics	$\langle 0.4, 0.9\rangle$	$\langle 0.2, 0.9\rangle$
12674	Sandy Dewitt	Marketing	$\langle 0.1, 0.8\rangle$	$\langle 0.3, 0.7\rangle$
56739	Bea Anthony	Accounting	$\langle 0.7, 0.6\rangle$	$\langle 0.2, 0.9\rangle$

FCLTY

F_ID	DEPT	EVALUATION	F_NAME	CI
12312	Marketing	$\langle 0.5, 0.7\rangle$	John Smith	$\langle 0, 1\rangle$
31897	Mathematics	$\langle 0.4, 0.9\rangle$	Mary Lee	$\langle 0.2, 0.9\rangle$
12674	Marketing	$\langle 0.1, 0.8\rangle$	Sandy Dewitt	$\langle 0.3, 0.7\rangle$
56739	Accounting	$\langle 0.7, 0.6\rangle$	Bea Anthony	$\langle 0.2, 0.9\rangle$

Note that the tuple 1 in the FACULTY ciset relation and the tuple 1 in the FCLTY ciset relation have the same information content that can be stated in plain English as "According to various sources, John Smith is in Marketing department and he has faculty identification number 12312 is 100% certain and 0% uncertain. Further, his student evaluation is $\langle 0.5, 0.7\rangle$". The second tuple in the FACULTY ciset relation and the second tuple in the FCLTY ciset relation convey the fact that "According to various sources, Mary Lee is in Mathematics department and has faculty identification number 31897 is 90% certain and 20% uncertain. Her student evaluation is $\langle 0.4, 0.9\rangle$" Thus each tuple conveys the same semantic interpretation in both ciset relations and hence both ciset relations are identical in all respects from the practical perspective as well.

Property 2: The order of tuples in a ciset relation is immaterial

A ciset relation is considered as a set of mappings. Therefore, the order of tuples in a ciset relation is also immaterial, from a theoretical point of view. We now proceed to show that the order of tuples in a ciset relation is immaterial, from a practical point of view. For example, the following two ciset relations are the same.

Consider two ciset relations FACULTY and FTY on the ciset relational scheme F given below.

FACULTY

F_ID	F_NAME	DEPT	EVALUATION	CI
12312	John Smith	Marketing	$\langle 0.5, 0.7\rangle$	$\langle 0, 1\rangle$
31897	Mary Lee	Mathematics	$\langle 0.4, 0.9\rangle$	$\langle 0.2, 0.9\rangle$
12674	Sandy Dewitt	Marketing	$\langle 0.1, 0.8\rangle$	$\langle 0.3, 0.7\rangle$
56739	Bea Anthony	Accounting	$\langle 0.7, 0.6\rangle$	$\langle 0.2, 0.9\rangle$

FTY

F_ID	F_NAME	DEPT	EVALUATION	CI
12674	Sandy Dewitt	Marketing	$\langle 0.1, 0.8\rangle$	$\langle 0.3, 0.7\rangle$
56739	Bea Anthony	Accounting	$\langle 0.7, 0.6\rangle$	$\langle 0.2, 0.9\rangle$
12312	John Smith	Marketing	$\langle 0.5, 0.7\rangle$	$\langle 0, 1\rangle$
31897	Mary Lee	Mathematics	$\langle 0.4, 0.9\rangle$	$\langle 0.2, 0.9\rangle$

The tuple 1 of FACULTY ciset relation appears as the tuple 3 of FTY ciset relation. The information content of those two tuples is the same and can be stated in plain English as "According to various sources, John Smith is in Marketing department and he has faculty identification number 12312 is 100% certain and 0% uncertain. Further, his student evaluation is $\langle 0.5, 0.7\rangle$". The tuple 2 of FACULTY ciset relation appears as tuple 4 of FTY ciset relation and so on. Thus every tuple of FACULTY ciset relation appears as a tuple of FTY ciset relation and vice-versa, except for their order of appearances. Since a tuple conveys the same semantic interpretation in both ciset relations, ciset relations FACULTY and FTY are identical in all respects from the practical perspective as well.

Property 3: If r is a ciset relation then no two tuples of $\mathcal{R}(r)$ are identical.

A ciset relation is considered as a set of mappings, each mapping corresponds to a tuple in the ciset relation. From the definition of a set, no two members of a set are identical. Therefore, in a ciset relation, no two mappings are identical. Note that values of attributes that are not relational can not be predicted and keep on changing depending upon the information received, the mappings can remain non-identical if and only if no two tuples are identical in $\mathcal{R}(r)$.

By keeping two tuples that are identical in $\mathcal{R}(r)$, we end up fragmenting the data. The main aim of this work is to avoid such a situation. For example, consider the following table.

FACULT

F_ID	F_NAME	DEPT	EVALUATION	CI
12312	John Smith	Marketing	$\langle 0.5, 0.7\rangle$	$\langle 0, 1\rangle$
31897	Mary Lee	Mathematics	$\langle 0.3, 0.9\rangle$	$\langle 0.2, 0.7\rangle$
31897	Mary Lee	Mathematics	$\langle 0.4, 0.6\rangle$	$\langle 0.1, 0.9\rangle$
12674	Sandy Dewitt	Marketing	$\langle 0.1, 0.8\rangle$	$\langle 0.3, 0.7\rangle$
56739	Bea Anthony	Accounting	$\langle 0.7, 0.6\rangle$	$\langle 0.2, 0.9\rangle$

In the above table there are two rows with identical values for F_ID, F_NAME and DEPT. In other words, the property 3 is violated. Therefore, above table is not considered as a ciset relation. Note that you can combine the information in row 2 and row 3 together by taking ciset union operation on columns labeled EVALUATION and CI; and thus obtain a single row. The resulting table

FACULTY

F_ID	F_NAME	DEPT	EVALUATION	CI
12312	John Smith	Marketing	$\langle 0.5, 0.7\rangle$	$\langle 0, 1\rangle$
31897	Mary Lee	Mathematics	$\langle 0.3, 0.9\rangle$	$\langle 0.1, 0.9\rangle$
12674	Sandy Dewitt	Marketing	$\langle 0.1, 0.8\rangle$	$\langle 0.3, 0.7\rangle$
56739	Bea Anthony	Accounting	$\langle 0.7, 0.6\rangle$	$\langle 0.2, 0.9\rangle$

is a ciset relation.

Property 4: A ciset relation is in first normal form

A ciset relation is considered as a set of mappings. Each mapping corresponds to one tuple in the ciset relation. From the definition of a ciset relation, each mapping assigns a single value to each of the attributes. This property is known as first normal form.

A ciset relation is in *First Normal Form (1NF)*, if every attribute value is an atomic value and not a list of values or a set of values. *Note that a confidence index is treated as a single value similar to a complex number.* We always keep the ciset relation in the first normal form. That is, no tuple in the ciset relation has more than one value for each of its attributes. Note that this requirement is implicit in our definition of a ciset relation.

Example 3.1.2 *Following is a table of employees and cities they have lived before.*

LVD_IN

EMPLOYEE_NAME	*CITY*	*CI*
John Smith	*New York, Chicago, Los Angeles*	$\langle 0.3, 0.7\rangle$
Mary Lee	*New York, San Francisco*	$\langle 0.5, 0.9\rangle$
Sandy Dewitt	*Boston, Paris, New Delhi*	$\langle 0.1, 0.8\rangle$
Bea Anthony	*Kansas City*	$\langle 0.2, 0.6\rangle$

This table is not in first normal form. In the case of John Smith, the CITY attribute has multiple values. So is the case with Mary Lee and Sandy Dewitt. In other words, the DOM(CITY) = {{ New York, Chicago, Los Angeles}, { New York, San Francisco}, { Boston, Paris, New Delhi}, { Kansas City}}. Thus elements of DOM(CITY) are sets; and not atomic values. We must keep the table in first normal form as shown below:

LIVED_IN

EMPLOYEE_NAME	*CITY*	*CI*
John Smith	*New York*	$\langle 0.3, 0.7\rangle$
John Smith	*Chicago*	$\langle 0.5, 0.8\rangle$
John Smith	*Los Angeles*	$\langle 0.4, 0.9\rangle$
Mary Lee	*New York*	$\langle 0.5, 1\rangle$
Mary Lee	*San Francisco*	$\langle 0.7, 0.9\rangle$
Sandy Dewitt	*Boston*	$\langle 0.1, 0.9\rangle$
Sandy Dewitt	*Paris*	$\langle 0.4, 0.8\rangle$
Sandy Dewitt	*New Delhi*	$\langle 0.3, 1\rangle$
Bea Anthony	*Kansas City*	$\langle 0.2, 0.6\rangle$

Note that DOM(CITY) is no longer a set of sets.

3.2 Integrity Constraints

superkey

Consider the Table 3.3.

The attribute F_ID stands for faculty identification number. This number is unique for each faculty. No two faculty members can have the same faculty identification number. We call such an attribute a superkey. The attribute F_NAME however is not unique. Even though at this time, there is only one John Smith among the faculty members, there is no reason to believe that in the future there will not be another faculty member with the same name. Thus F_NAME can not be a superkey. The remaining attributes can not be superkeys either. Thus the following ciset relation FACULTY and the relation $\mathcal{R}$(FACULTY)

FACULTY

F_ID	F_NAME	DEPT	EVALUATION	CI
12312	John Smith	Marketing	$\langle 0.5, 0.7\rangle$	$\langle 0, 1\rangle$
31897	Mary Lee	Mathematics	$\langle 0.4, 0.9\rangle$	$\langle 0.2, 0.9\rangle$
12674	Sandy Dewitt	Marketing	$\langle 0.1, 0.8\rangle$	$\langle 0.3, 0.7\rangle$
56739	Bea Anthony	Accounting	$\langle 0.7, 0.6\rangle$	$\langle 0.2, 0.9\rangle$

$\mathcal{R}$(FACULTY)

F_ID	F_NAME	DEPT
12312	John Smith	Marketing
31897	Mary Lee	Mathematics
12674	Sandy Dewitt	Marketing
56739	Bea Anthony	Accounting

has F_ID as a superkey.

Let $r = \{t_1, \ldots, t_m\}$ be a ciset relation on a ciset relational scheme R. A nonempty subset of relational attributes X of $\mathcal{R}(R)$ is called a *superkey*, if for any two distinct tuples $t_i, t_j \in r, t_i(X) \neq t_j(X)$.

For example consider the ciset relational scheme
$F = \{F\_ID, F\_NAME, DEPT, CONFIDENCE, CI\}$. Now
FACULTY is a ciset relation on F. Thus

$FACULTY = \{(12312, John\ Smith, Marketing, \langle 0.5, 0.7\rangle, \langle 0, 1\rangle),$
$(31897, Mary\ Lee, Mathematics, \langle 0.4, 0.9\rangle, \langle 0.2, 0.9\rangle),$
$(12674, Sandy\ Dewitt, Marketing, \langle 0.1, 0.8\rangle, \langle 0.3, 0.7\rangle),$
$(56739, Bea\ Anthony, Accounting, \langle 0.7, 0.6\rangle, \langle 0.2, 0.9\rangle)\}.$

Further,
$\mathcal{R}(FACULTY) = \{(12312, John\ Smith, Marketing),$
$(31897, Mary\ Lee, Mathematics),$
$(12674, Sandy\ Dewitt, Marketing),$
$(56739, Bea\ Anthony, Accounting)\}.$

Now the set of all superkeys of $FACULTY$ is
$\{\{F\_ID\}, \{F\_ID, F\_NAME\}, \{F\_ID, DEPT\},$
$\{F\_ID, F\_NAME, DEPT\}\}$
and the set of all superkeys of $\mathcal{R}(FACULTY)$ is
$\{\{F\_ID\}, \{F\_ID, F\_NAME\}, \{F\_ID, DEPT\},$
$\{F\_ID, F\_NAME, DEPT\}\}$
Thus the set of all superkeys of $FACULTY$ is the same as the set of all superkeys of $\mathcal{R}(FACULTY)$.

The following result is quite obvious, so we state it without proof.

Theorem 3.2.1 *Let r be a ciset relation on a ciset relational scheme R and let $s = \mathcal{R}(r)$. Then r and s have the same set of superkeys.*

Properties of a superkey

Property 1: Every ciset relation has a superkey

This is an immediate consequence of the Property 3 of a ciset relation. According to Property 3 of a ciset relation, if r is a ciset relation then no two tuples of $\mathcal{R}(r)$ are identical. Therefore, if r is a ciset relation on a ciset relational scheme R, then $\mathcal{R}(R)$ is a superkey. Note that this result can also be obtained from Theorem 3.2.1 and Property 1 of a superkey of a relational database.

Property 2: If X is a superkey, so is any super set of X.

This property needs to be explained in more detailed fashion. Let r be a ciset relation on a ciset relational scheme R and let X be a superkey of r. The Property 2 may be stated in more detail as follows.

Let X be a superkey on a ciset relational scheme R and Y be such that $X \subseteq Y \subseteq \mathcal{R}(R)$. Then Y is a superkey.

Assume that X is a superkey. Therefore, for any two tuples t, t' of $r, t(X)$ and $t'(X)$ are not identical. Now Y has all the attributes of X and as a consequence for any two tuples t, t' of $r, t(Y)$ and $t'(Y)$ are not identical. Further, $Y \subseteq \mathcal{R}(R)$. Thus Y is a superkey. Once again, observe that this result can also be obtained from Theorem 3.2.1 and Property 2 of a superkey of a relational database.

Example 3.2.2 *Recall that* $X = \{F\_ID\}$ *is a superkey of the ciset relation FACULTY first presented in Table 3.3. Let* $Y = \{F\_ID, F\_NAME\}$. *Note that* $X = \{F\_ID\} \subseteq Y = \{F\_ID, F\_NAME\} \subseteq \mathcal{R}(R)$
$= \{F\_ID, F\_NAME, DEPT\}$. *Property 2 states that*
$\{F\_ID, F\_NAME\}$ *is a superkey. This can be verified by observing the fact that all of the following tuples*
$t_1(Y) = (12312, Mary\ Smith)$,
$t_2(Y) = (31897, Mary\ Lee,)$,
$t_3(Y) = (12674, Sandy\ Dewitt)$,
$t_4(Y) = (56739, Bea\ Anthony)$,
are distinct.

Property 3: Every superkey contains a minimal superkey.

The term minimal warrants some explanation. Starting with a superkey X, it is possible to remove one of ciset relational attributes from X to obtain a proper subset Y of X. Verify whether or not Y is a superkey. If Y is not a superkey, we ignore Y and start with X again. However, if Y is a superkey, we replace X with Y and repeat the process again. This process will end with a set X of ciset relational attributes such that X is a superkey and no proper subset of X is a superkey. The superkey X obtained by the above process is called a minimal superkey. In other words, a superkey X is a minimal superkey if no proper subset of X is a superkey.

It may be observed that this result can also be obtained from Theorem 3.2.1 and Property 3 of a superkey in relational database. The algorithm to compute a minimal superkey is the same as the one presented in Chapter 2.

Candidate key

In the previous subsection, we have noticed that a ciset relation may have many superkeys. In fact, we have seen that a ciset relation may have many minimal superkeys. Minimal superkeys play an important role in database design and they are called candidate keys.

A *candidate key* is an attribute or a combination of attributes that uniquely identifies each tuple of a ciset relation and is minimal.

Let r be a ciset relation on a ciset relational scheme R. A nonempty subset of relational attributes X of R is called a candidate key, if for any two distinct tuples $t, t' \in r, t(X)$ is different from $t'(X)$ and for every proper subset of X' of X, there exists at least two distinct tuples $t, t' \in r, t(X')$ is the same as $t'(X')$.

In terms of superkeys, a candidate key can be formally defined as follows.

Let r be a ciset relation on relational scheme R. A nonempty subset of relational attributes X of R is called a candidate key, if X is a superkey and no proper subset of X is a superkey. Due to Theorem 3.2.1, we have the following.

Theorem 3.2.3 *Let r be a ciset relation on a ciset relational scheme R. Let $s = \mathcal{R}(r)$. Then r and s have the same set of candidate keys.*

Primary key

In the previous subsection, we have observed that a ciset relation may have many candidate keys. One of the candidate keys is selected as the primary key of the ciset relation. In is important to consider various issues in our selection process and is the same as in the case of selection of primary key in traditional relational database. Reader is referred to Chapter 2 for a more detailed description.

Once a primary key is selected, we insist that primary key value cannot be null. That is, all attributes involved in the primary key must have a valid data entry. From the definition of the primary key, it follows that primary key value cannot be repeated. Thus we have the **first data integrity rule**, known as, entity integrity rule.

A primary key value cannot be repeated and it cannot be a null value.

The purpose of this integrity rule is to ensure that every tuple in a ciset relation has a unique identifier.

Every ciset relation in a database must obey the entity integrity rule.

Foreign key

Another level of data integrity is enforced through foreign keys. Since the foreign key concept is the same as the foreign key concept in traditional database. The reader is referred to Chapter 2 for a more detailed description.

The second data integrity rule, also known as, **referential integrity rule** can be stated as follows.

A foreign key entry can be null or an entry that matches a primary key value of the referencing table.

The purpose of this integrity rule is to ensure that every entity referred in fact exists in the referencing table. We allow null values, as a practical consideration.

Every ciset relation in a database must obey the referential integrity rule.

4 THE CISET RELATIONAL ALGEBRA

In this chapter we introduce the ciset relational operations. These operations extend traditional relational operations. We begin our discussion with set theoretic operations.

4.1 Union, Intersection and Difference

Two ciset relational schemes are said to be *union compatible*, if they have the same attribute characteristics. More formally, we define two ciset relational schemes $R = \{A_1, A_2, \ldots, A_n\}$ and $S = \{B_1, B_2, \ldots, B_m\}$ to be union compatible if the following conditions hold:

1. $n = m$ or the number of attributes is the same;
2. $DOM(A_i) = DOM(B_i)$, for $i = 1, 2, \ldots, n$.

Let REL_R and REL_S be any two ciset relations on ciset relational schemes R and S respectively. Relational operations, union, intersection and difference on REL_R and REL_S are defined, if and only if R and S are union compatible.

For the rest of this section, assume that REL_R and REL_S are two ciset relations on ciset relational schemes R and S respectively and R and S are union compatible. Further, without loss of generality assume that $R = S = \{A_1, A_2, \ldots, A_n\}$, where $\{A_1, A_2, \ldots, A_k\}$ are relational attributes and $\{A_{k+1}, A_{k+2}, \ldots, A_n\}$ are non-relational attributes. In other words, $DOM(A_i)$, for $i = k+1, k+2, \ldots, n$ are subsets of $\mathfrak{C}$. Each tuple of REL_R and REL_S can be represented as $(t, \alpha_{k+1}(t), \alpha_{k+2}(t), \ldots, \alpha_n(t))$ where $t \in DOM(A_1) \times DOM(A_2) \times \ldots \times DOM(A_k)$ and $\alpha_i(t) \in DOM(A_i)$ for $i = k+1, k+2, \ldots, n$.

Union

The *union* is a binary operation. When applied to two union compatible ciset relations, it yields a new ciset relation that is union compatible to both of them. The union of REL_R and REL_S is obtained by combining the rows of both REL_R and REL_S. We use the notation $REL\_R \cup REL\_S$ to denote the union of REL_R and REL_S.

In other words, if
$REL\_R = \{(t_i, \alpha_{i,k+1}(t), \alpha_{i,k+2}(t), \ldots, \alpha_{i,n}(t)) \mid i \in I\}$
and
$REL\_S = \{(t_j, \alpha_{j,k+1}(t), \alpha_{j,k+2}(t), \ldots, \alpha_{j,n}(t)) \mid j \in J\}$,
then $REL\_R \cup REL\_S =$
$\{(t_i, \alpha_{i,k+1}(t) \cup \alpha_{j,k+1}(t), \alpha_{i,k+2}(t) \cup \alpha_{j,k+2}(t), \ldots, \alpha_{i,n}(t)$
$\cup \alpha_{j,n}(t)) \mid i \in I, j \in J, t_i = t_j\} \cup \{(t_i, \alpha_{i,k+1}(t),$
$\alpha_{i,k+2}(t), \ldots, \alpha_{i,n}(t)) \mid i \in I, \nexists j \in J$ such that $t_i = t_j\}$
$\cup \{(t_j, \alpha_{j,k+1}(t), \alpha_{j,k+2}(t), \ldots, \alpha_{j,n}(t)) \mid j \in J, \nexists i \in I$
such that $t_i = t_j\}$.

Example 4.1.1 *We use the following two union compatible ciset relations to illustrate the union operation.*

REL_R

A	B	C	CI
a_1	b_1	c_1	$\langle 0.3, 0.7 \rangle$
a_1	b_1	c_2	$\langle 0.6, 0.2 \rangle$
a_3	b_3	c_3	$\langle 0, 1 \rangle$
a_4	b_4	c_4	$\langle 1, 1 \rangle$
a_5	b_5	c_5	$\langle 0, 0 \rangle$
a_2	b_2	c_1	$\langle 0.4, 0 \rangle$
a_2	b_1	c_2	$\langle 0.3, 0 \rangle$

REL_S

A	B	C	CI
a_1	b_2	c_2	$\langle 0.2, 0.5 \rangle$
a_2	b_2	c_1	$\langle 0.1, 0.8 \rangle$
a_3	b_3	c_3	$\langle 1, 1 \rangle$
a_4	b_4	c_4	$\langle 0, 0 \rangle$
a_5	b_5	c_5	$\langle 0, 1 \rangle$
a_1	b_1	c_2	$\langle 0, 0.8 \rangle$

$REL\_R \cup REL\_S$

A	B	C	CI
a_1	b_1	c_1	$\langle 0.3, 0.7\rangle$
a_1	b_1	c_2	$\langle 0, 0.8\rangle$
a_2	b_2	c_1	$\langle 0.1, 0.8\rangle$
a_3	b_3	c_3	$\langle 0, 1\rangle$
a_4	b_4	c_4	$\langle 0, 1\rangle$
a_5	b_5	c_5	$\langle 0, 1\rangle$
a_2	b_1	c_2	$\langle 0.3, 0\rangle$
a_1	b_2	c_2	$\langle 0.2, 0.5\rangle$

Intersection

The intersection is a binary operation. When applied to two union compatible ciset relations, it yields a new ciset relation that is union compatible to both of them. The intersection of REL_R and REL_S is obtained by taking rows common to both REL_R and REL_S. The notation $REL\_R \cap REL\_S$ is used to denote the intersection of REL_R and REL_S.

In other words, if $REL\_R = \{(t_i, \alpha_{i,k+1}(t), \alpha_{i,k+2}(t), \ldots, \alpha_{i,n}(t)) \mid i \in I\}$ and $REL\_S = \{(t_j, \alpha_{j,k+1}(t), \alpha_{j,k+2}(t), \ldots, \alpha_{j,n}(t)) \mid j \in J\}$, then $REL\_R \cap REL\_S = \{(t_i, \alpha_{i,k+1}(t) \cap \alpha_{j,k+1}(t), \alpha_{i,k+2}(t) \cap \alpha_{j,k+2}(t), \ldots, \alpha_{i,n}(t) \cap \alpha_{j,n}(t)) \mid i \in I, j \in J, t_i = t_j\}$.

Example 4.1.2 *Let REL_R and REL_S be as in Example 4.1.1.*

$REL\_R \cap REL\_S$

A	B	C	CI
a_1	b_1	c_2	$\langle 0.6, 0.2\rangle$
a_3	b_3	c_3	$\langle 1, 1\rangle$
a_5	b_5	c_5	$\langle 0, 0\rangle$
a_2	b_2	c_1	$\langle 0.4, 0\rangle$

Note that $(a_4, b_4, c_4, \langle 1, 0\rangle)$ is not shown in $REL\_R \cap REL\_S$.

Difference

The *difference* is a binary operation. When applied to two union compatible ciset relations, it yields a new ciset relation that is union compatible to both of them. We use the notation $REL\_R - REL\_S$ to denote the difference of REL_R from REL_S. Note that this definition is the same as the ciset difference operation introduced in Chapter 1. In other words, let $REL\_R = \{(t_i, \alpha_{i,k+1}(t), \alpha_{i,k+2}(t), \ldots, \alpha_{i,n}(t)) \mid i \in I\}$ and $REL\_S = \{(t_j, \alpha_{j,k+1}(t), \alpha_{j,k+2}(t), \ldots, \alpha_{j,n}(t)) \mid j \in J\}$.Then $REL\_R - REL\_S =$

$\{(t_i, \alpha_{i,k+1}(t), \alpha_{i,k+2}(t), \ldots, \alpha_{i,n}(t)) \mid i \in I, \nexists j \in J \text{ such that } t_i = t_j\} \cup$
$\{(t_i, \alpha_{i,k+1}(t) - \alpha_{j,k+1}(t), \alpha_{i,k+2}(t) - \alpha_{j,k+2}(t), \ldots, \alpha_{i,n}(t) - \alpha_{j,n}(t)) \mid$
$i \in I, j \in J, t_i = t_j\}$.

Example 4.1.3 *We use the following two union compatible ciset relations to illustrate the difference.*

REL_R

A	B	C	CI
a_1	b_1	c_1	$\langle 0.3, 0.7\rangle$
a_1	b_1	c_2	$\langle 0.6, 0.2\rangle$
a_3	b_3	c_3	$\langle 0, 1\rangle$
a_4	b_4	c_4	$\langle 1, 1\rangle$
a_5	b_5	c_5	$\langle 0, 0\rangle$
a_6	b_6	c_6	$\langle 0, 1\rangle$
a_7	b_7	c_7	$\langle 1, 1\rangle$
a_8	b_8	c_8	$\langle 0, 0\rangle$
a_9	b_9	c_9	$\langle 0, 1\rangle$
a_{10}	b_{10}	c_{10}	$\langle 1, 1\rangle$
a_{11}	b_{11}	c_{11}	$\langle 0, 0\rangle$
a_2	b_2	c_1	$\langle 0.4, 0\rangle$
a_2	b_1	c_2	$\langle 0.3, 0\rangle$

REL_S

A	B	C	CI
a_1	b_2	c_2	$\langle 0.2, 0.5\rangle$
a_2	b_2	c_1	$\langle 0.1, 0.8\rangle$
a_3	b_3	c_3	$\langle 0, 1\rangle$
a_4	b_4	c_4	$\langle 0, 1\rangle$
a_5	b_5	c_5	$\langle 0, 1\rangle$
a_6	b_6	c_6	$\langle 1, 1\rangle$
a_7	b_7	c_7	$\langle 1, 1\rangle$
a_8	b_8	c_8	$\langle 1, 1\rangle$
a_9	b_9	c_9	$\langle 0, 0\rangle$
a_{10}	b_{10}	c_{10}	$\langle 0, 0\rangle$
a_{11}	b_{11}	c_{11}	$\langle 0, 0\rangle$
a_1	b_1	c_2	$\langle 0, 0.8\rangle$

REL_R − REL_S

A	B	C	CI
a_1	b_1	c_1	$\langle 0.3, 0.7\rangle$
a_1	b_1	c_2	$\langle 0.8, 0\rangle$
a_6	b_6	c_6	$\langle 1, 1\rangle$
a_7	b_7	c_7	$\langle 1, 1\rangle$
a_8	b_8	c_8	$\langle 1, 0\rangle$
a_9	b_9	c_9	$\langle 0, 0\rangle$
a_{11}	b_{11}	c_{11}	$\langle 0, 0\rangle$
a_2	b_2	c_1	$\langle 0.4, 0\rangle$
a_2	b_1	c_2	$\langle 0.3, 0\rangle$

In the next example, we show that ciset relational operations are in fact a true extension of relational operations.

Example 4.1.4 *Consider the following two union compatible relations on a relational scheme* $\{A, B, C\}$.

REL_R

A	B	C
a_1	b_1	c_1
a_1	b_1	c_2
a_2	b_2	c_1
a_2	b_1	c_2

REL_S

A	B	C
a_1	b_2	c_2
a_2	b_2	c_1
a_1	b_1	c_2

First, let us covert them to equivalent ciset relations by adding one more attribute, CI, to REL_R and REL_S; and thus obtain F_{REL_R} and F_{REL_S} respectively.

F_{REL_R}

A	B	C	CI
a_1	b_1	c_1	$\langle 0,1\rangle$
a_1	b_1	c_2	$\langle 0,1\rangle$
a_2	b_2	c_1	$\langle 0,1\rangle$
a_2	b_1	c_2	$\langle 0,1\rangle$

F_{REL_S}

A	B	C	CI
a_1	b_2	c_2	$\langle 0,1\rangle$
a_2	b_2	c_1	$\langle 0,1\rangle$
a_1	b_1	c_2	$\langle 0,1\rangle$

Now $F_{REL\_R} \cup F_{REL\_S}, F_{REL\_R} \cap F_{REL\_S}$ and $F_{REL\_R} - F_{REL\_S}$ are given by

$F_{REL\_R} \cup F_{REL\_S}$

A	B	C	CI
a_1	b_1	c_1	$\langle 0,1\rangle$
a_1	b_1	c_2	$\langle 0,1\rangle$
a_2	b_2	c_1	$\langle 0,1\rangle$
a_2	b_1	c_2	$\langle 0,1\rangle$
a_1	b_2	c_2	$\langle 0,1\rangle$

$F_{REL\_R} \cap F_{REL\_S}$

A	B	C	CI
a_2	b_2	c_1	$\langle 0,1\rangle$
a_1	b_1	c_2	$\langle 0,1\rangle$

$F_{REL\_R} - F_{REL\_S}$

A	B	C	CI
a_1	b_1	c_1	$\langle 0,1\rangle$
a_2	b_1	c_2	$\langle 0,1\rangle$

Thus we have following three relations

$(F_{REL\_R} \cup F_{REL\_S})^1$

A	B	C
a_1	b_1	c_1
a_1	b_1	c_2
a_2	b_2	c_1
a_2	b_1	c_2
a_1	b_2	c_2

$(F_{REL\_R} \cap F_{REL\_S})^1$

A	B	C
a_2	b_2	c_1
a_1	b_1	c_2

$(F_{REL\_R} - F_{REL\_S})^1$

A	B	C
a_1	b_1	c_1
a_2	b_1	c_2

It may be noted that

$REL\_R \cup REL\_S = (F_{REL\_R} \cup F_{REL\_S})^1$,

$REL\_R \cap REL\_S = (F_{REL\_R} \cap F_{REL\_S})^1$ and

$REL\_R - REL\_S = (F_{REL\_R} - F_{REL\_S})^1$.

Theorem 4.1.5 *Let REL_R, REL_S any two relations on a relational scheme R. Then the following diagrams commute.*

$$\begin{array}{ccc} (REL\_R, REL\_S) & \xrightarrow{F} & (F_{REL\_R}, F_{REL\_S}) \\ \downarrow \cup & & \downarrow \cup \\ REL\_R \cup REL\_S & \xleftarrow{(\,)^1} & F_{REL\_R} \cup F_{REL\_S} \end{array}$$

$$\begin{array}{ccc} (REL\_R, REL\_S) & \xrightarrow{F} & (F_{REL\_R}, F_{REL\_S}) \\ \downarrow \cap & & \downarrow \cap \\ REL\_R \cap REL\_S & \xleftarrow{(\,)^1} & F_{REL\_R} \cap F_{REL\_S} \end{array}$$

$$\begin{array}{ccc} (REL\_R, REL\_S) & \xrightarrow{F} & (F_{REL\_R}, F_{REL\_S}) \\ \downarrow - & & \downarrow - \\ REL\_R - REL\_S & \xleftarrow{(\,)^1} & F_{REL\_R} - F_{REL\_S} \end{array}$$

Proof. Result follows from Theorem 1.3.14. ■

The above theorem is a very powerful result. It simply states, given two union compatible relations REL_R and REL_S, the basic three relational operations union, intersection and difference can either be computed as such or REL_R and REL_S can be converted into equivalent ciset relations and the corresponding ciset relational operations can be carried out and converted back to relations without any loss of information. Since this result holds for three basic operations, any operation defined using these basic operations shall also inherit this property.

4.2 Select and Project

Select

The *select* is a unary operation. When applied to a ciset relation, it yields a new union compatible ciset relation. The select operation produces a new ciset relation by selecting tuples that satisfies a given predicate. In addition to attributes, predicate can also contain two key words UL and LL where UL refers to the upper index of the confidence index and LL refers to the lower index of the confidence index. We use σ to denote the select operation.

To be more specific, let REL_R be a ciset relation on a relational scheme $R = \{A, B, C, D\}$. For $a \in DOM(A), \sigma_{A\,=\,a}REL\_R$ is a select operation on REL_R that will create a new ciset relation having all tuples of REL_R with attribute A equal to a. Further if $b \in DOM(B)$ then selection operation $\sigma_{(A\,=\,a)\wedge(B\,=\,b)}REL\_R$ is a unary operation on REL_R that creates a new ciset relation having all tuples of REL_R with attribute A equal to a and attribute $B = b$. If $DOM(D) = \mathfrak{C}$ and $\langle\alpha, \beta\rangle$ is a confidence index, then

$$\sigma_{(A\,=\,a)\wedge(D.LL\,<\,\alpha)\wedge(D.UL\,\geq\beta)}REL\_R$$

is a select operation on REL_R that will create a new ciset relation having all tuples of REL_R with attribute A equal to a and D value is such that lower index is less then α and upper index is greater than or equal to β.

Example 4.2.1 *Let REL_R be as shown below.*

REL_R

A	B	C	D
a_1	b_1	c_1	$\langle 0.3, 0.7 \rangle$
a_1	b_1	c_2	$\langle 0.6, 0.2 \rangle$
a_3	b_2	c_1	$\langle 0.8, 0 \rangle$
a_1	b_1	c_3	$\langle 0.4, 0.9 \rangle$

Then

$\sigma_{A=a_1} REL\_R$

A	B	C	D
a_1	b_1	c_1	$\langle 0.3, 0.7 \rangle$
a_1	b_1	c_2	$\langle 0.6, 0.2 \rangle$
a_1	b_1	c_3	$\langle 0.4, 0.9 \rangle$

and

$\sigma_{(A=a_1) \wedge (D.LL<0.5) \wedge (D.UL \geq 0.7)} REL\_R$

A	B	C	D
a_1	b_1	c_1	$\langle 0.3, 0.7 \rangle$
a_1	b_1	c_2	$\langle 0.4, 0.9 \rangle$

The proof of the following theorem is quite clear. Hence we state it without any proof.

Theorem 4.2.2 *Let REL_R be a relation on a relational scheme R. Then the following diagram commute.*

$$\begin{array}{ccc} REL\_R & \xrightarrow{F} & F_{REL\_R} \\ \downarrow \sigma & & \downarrow \sigma \\ \sigma_{criteria} REL\_R & \xleftarrow{(\,)^1} & \sigma_{criteria} F_{REL\_R} \end{array}$$

■

Project

The project is a unary operation. The project operation discards unwanted attributes. We use the notation Π to denote the project operation.

Let REL_R be a ciset relation on a ciset relational scheme $R = \{A, B, C, D\}$. For $X \subseteq R, \Pi_X REL\_A$ is defined as a ciset relation on a ciset relational scheme X such that t is a tuple in REL_R if and only if $t(X)$ is a tuple in $\Pi_X REL\_A$. For example, if $X = \{A, C\}$, we have the following:

$\Pi_{\{A,C\}} REL\_A$

A	C
a_1	c_1
a_1	c_2
a_1	c_3
a_3	c_1

and further, if $X = \{A, C, D\}$, we have the following:
$\Pi_{\{A,C,D\}} REL\_A$

A	C	D
a_1	c_1	$\langle 0.3, 0.7 \rangle$
a_1	c_2	$\langle 0.6, 0.9 \rangle$
a_3	c_1	$\langle 0.8, 0 \rangle$

Up to this point in our discussion, we used the terminology $\langle 0, 1 \rangle$–cut. Clearly, in the context of ciset relational operations, it may be noted that $\langle 0, 1 \rangle$–cut on the attribute D in the above ciset relational scheme is equivalent to $\Pi_{\{A,B,C\}} \sigma_{(D.LL=0) \wedge (D.UL=1)}$.

The proof of the following theorem is quite clear. Hence we state it without any proof.

Theorem 4.2.3 *Let REL_R be a relation on a relational scheme R. Then the following diagram commute.*

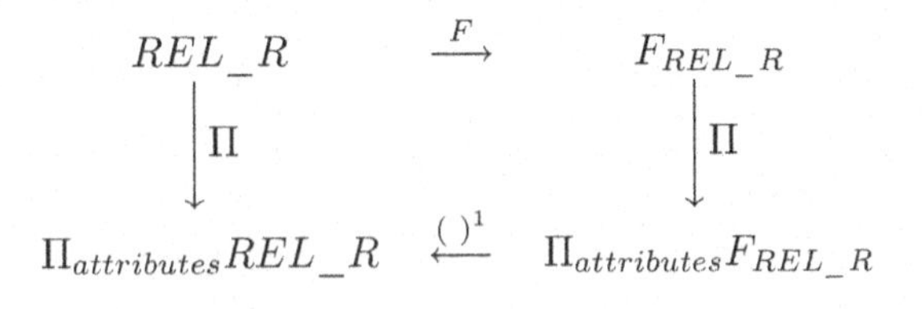

$$\begin{array}{ccc} REL\_R & \xrightarrow{F} & F_{REL\_R} \\ \downarrow \Pi & & \downarrow \Pi \\ \Pi_{attributes} REL\_R & \xleftarrow{(\,)^1} & \Pi_{attributes} F_{REL\_R} \end{array}$$

■

4.3 Product and Join

Product

The *product* is a binary operation. Let REL_R and REL_T be two ciset relations on ciset relational schemes $R = \{A_1, A_2, \ldots, A_n, C_1, \ldots, C_k\}$ and $T = \{B_1, B_2, \ldots, B_m, C_1, \ldots, C_k\}$ respectively where $\{A_1, A_2, \ldots, A_n\}$ and

$\{B_1, B_2, \ldots, B_m\}$ are sets of ciset relational attributes and $\{C_1, C_2, \ldots, C_k\}$ is the set of attributes common to R and T such that $DOM(C_i) = \mathfrak{C}$. It must be noted that there may or may not exists common relational attributes between the sets of attributes $\{A_1, A_2, \ldots, A_n\}$ and $\{B_1, B_2, \ldots, B_m\}$. Then the product of REL_R and REL_T is a ciset relation on the scheme $\{A_1, A_2, \ldots, A_n, B_1, B_2, \ldots, B_m, C_1, \ldots, C_k\}$. The product produces a list of all possible pairs of tuple from two ciset relations. Therefore, if one ciset relation has five rows and the other has 10 rows, the ciset relation obtained by taking their product has at most 50 tuples. We use the notation $\times$ for the product operation.

In other words, let $REL\_R = \{(t_i, \alpha_{i,1}(t), \alpha_{i,2}(t), \ldots, \alpha_{i,k}(t)) \mid i \in I\}$ and $REL\_S = \{(t_j, \alpha_{j,1}(t), \alpha_{j,2}(t), \ldots, \alpha_{j,k}(t)) \mid j \in J\}$.

Then $REL\_R \times REL\_S$
$= \{(t_i, t_j, \alpha_{i,1}(t) \cap \alpha_{j,1}(t), \alpha_{i,2}(t) \cap \alpha_{j,2}(t), \ldots, \alpha_{i,k}(t) \cap \alpha_{j,k}(t)) \mid i \in I, j \in J\}$.

Let REL_R be a ciset relation on $R = \{A, B, C, F\}$ and REL_T be a ciset relation on $T = \{D, E, F\}$.

REL_R

A	B	C	F
a_1	b_1	c_1	$\langle 0.8, 0.4\rangle$
a_1	b_1	c_2	$\langle 1, 1\rangle$
a_2	b_2	c_1	$\langle 0, 0\rangle$
a_2	b_1	c_2	$\langle 0, 1\rangle$

REL_T

D	E	F
d_1	e_2	$\langle 0.7, 0.8\rangle$
d_2	e_2	$\langle 0, 0.9\rangle$
d_1	e_1	$\langle 0.6, 1\rangle$

Now $REL\_R \times REL\_T$ is a ciset relation on $\{A, B, C, D, E, F\}$ and has 11 tuples. It may be noted that each tuple is produced by one tuple from REL_R and one tuple from REL_T.

$REL\_R \times REL\_T$

A	B	C	D	E	F
a_1	b_1	c_1	d_1	e_2	$\langle 0.8, 0.4\rangle$
a_1	b_1	c_2	d_1	e_2	$\langle 1, 0.8\rangle$
a_2	b_2	c_1	d_1	e_2	$\langle 0.7, 0\rangle$
a_2	b_1	c_2	d_1	e_2	$\langle 0.7, 0.8\rangle$
a_1	b_1	c_1	d_2	e_2	$\langle 0.8, 0.4\rangle$
a_1	b_1	c_2	d_2	e_2	$\langle 1, 0.9\rangle$
a_2	b_2	c_1	d_2	e_2	$\langle 0, 0\rangle$
a_2	b_1	c_2	d_2	e_2	$\langle 0, 0.9\rangle$
a_1	b_1	c_1	d_1	e_1	$\langle 0.8, 0.4\rangle$
a_1	b_1	c_2	d_1	e_1	$\langle 1, 1\rangle$
a_2	b_2	c_1	d_1	e_1	$\langle 0.6, 0\rangle$
a_2	b_1	c_2	d_1	e_1	$\langle 0.6, 1\rangle$

Let REL_R be a relation on $R = \{A, B, C\}$ and REL_T be a relation on $T = \{D, E\}$.

REL_R

A	B	C
a_1	b_1	c_1
a_1	b_1	c_2
a_2	b_2	c_1
a_2	b_1	c_2

REL_T

D	E
d_1	e_2
d_2	e_2
d_1	e_1

Then

F_{REL_R}

A	B	C	CI
a_1	b_1	c_1	$\langle 0,1\rangle$
a_1	b_1	c_2	$\langle 0,1\rangle$
a_2	b_2	c_1	$\langle 0,1\rangle$
a_2	b_1	c_2	$\langle 0,1\rangle$

F_{REL_T}

D	E	CI
d_1	e_2	$\langle 0,1\rangle$
d_2	e_2	$\langle 0,1\rangle$
d_1	e_1	$\langle 0,1\rangle$

Now $F_{REL\_R} \times F_{REL\_T}$ is a ciset relation on $\{A, B, C, D, E, CI\}$ and has 12 tuples.

$F_{REL\_R} \times F_{REL\_T}$

A	B	C	D	E	CI
a_1	b_1	c_1	d_1	e_2	$\langle 0,1\rangle$
a_1	b_1	c_2	d_1	e_2	$\langle 0,1\rangle$
a_2	b_2	c_1	d_1	e_2	$\langle 0,1\rangle$
a_2	b_1	c_2	d_1	e_2	$\langle 0,1\rangle$
a_1	b_1	c_1	d_2	e_2	$\langle 0,1\rangle$
a_1	b_1	c_2	d_2	e_2	$\langle 0,1\rangle$
a_2	b_2	c_1	d_2	e_2	$\langle 0,1\rangle$
a_2	b_1	c_2	d_2	e_2	$\langle 0,1\rangle$
a_1	b_1	c_1	d_1	e_1	$\langle 0,1\rangle$
a_1	b_1	c_2	d_1	e_1	$\langle 0,1\rangle$
a_2	b_2	c_1	d_1	e_1	$\langle 0,1\rangle$
a_2	b_1	c_2	d_1	e_1	$\langle 0,1\rangle$

Therefore,
$(F_{REL\_R} \times F_{REL\_S})^1$

A	B	C	D	E
a_1	b_1	c_1	d_1	e_2
a_1	b_1	c_2	d_1	e_2
a_2	b_2	c_1	d_1	e_2
a_2	b_1	c_2	d_1	e_2
a_1	b_1	c_1	d_2	e_2
a_1	b_1	c_2	d_2	e_2
a_2	b_2	c_1	d_2	e_2
a_2	b_1	c_2	d_2	e_2
a_1	b_1	c_1	d_1	e_1
a_1	b_1	c_2	d_1	e_1
a_2	b_2	c_1	d_1	e_1
a_2	b_1	c_2	d_1	e_1

and thus
$REL\_R \times REL\_S = (F_{REL\_R} \times F_{REL\_S})^1$.

Theorem 4.3.1 *Let REL_R, REL_S two ciset relations on ciset relational schemes R and S respectively. Then the following diagram commute.*

$$\begin{array}{ccc} (REL\_R, REL\_S) & \xrightarrow{F} & (F_{REL\_R}, F_{REL\_S}) \\ \downarrow \times & & \downarrow \times \\ REL\_R \times REL\_S & \xleftarrow{(\,)^1} & F_{REL\_R} \times F_{REL\_S} \end{array}$$

Proof. Result follows from Theorem 1.3.14. ∎

Join

The *join* or *natural join* is a binary operation. Let REL_R and REL_T be two ciset relations on ciset relational schemes
$R = \{A_1, A_2, \ldots, A_n, D_1, D_2, \ldots, D_p, C_1, \ldots, C_k\}$
and
$T = \{B_1, B_2, \ldots, B_m, D_1, D_2, \ldots, D_p, C_1, \ldots, C_k\}$
respectively where $\{A_1, A_2, \ldots, A_n, D_1, D_2, \ldots, D_p\}$ and $\{B_1, B_2, \ldots, B_m, D_1, D_2, \ldots, D_p\}$ are sets of ciset relational attributes; $\{C_1, C_2, \ldots, C_k\}$ is the set of attributes common to R and T such that $DOM(C_i) = \mathfrak{C}$; and $\{D_1, D_2, \ldots, D_p\}$ is the set of relational attributes common to R and T. Join links tables REL_R and REL_T by selecting only rows with identical values in their common relational attribute(s) and produces a ciset relation on the scheme $\{A_1, A_2, \ldots, A_n, D_1, D_2, \ldots, D_p, B_1, B_2, \ldots, B_m, C_1, \ldots, C_k\}$. We use the

symbol $\bowtie$ for the natural join operation. In the special case where R and T have no common attributes, $REL\_R \bowtie REL\_T$ is the same as $REL\_R \times REL\_T$.

In other words, let $REL\_R = \{(t_i, s_i, \alpha_{i,1}(t), \alpha_{i,2}(t), \ldots, \alpha_{i,k}(t)) \mid i \in I\}$ and $REL\_S = \{(t_j, s_j, \alpha_{j,1}(t), \alpha_{j,2}(t), \ldots, \alpha_{j,k}(t)) \mid j \in J\}$. Here $t_i \in \Pi_{\{A_1,A_2,\ldots,A_n\}}REL\_R$, $s_i \in \Pi_{\{D_1,D_2,\ldots,D_p\}}REL\_R$, $t_j \in \Pi_{\{B_1,B_2,\ldots,B_m\}}REL\_S$, and $s_j \in \Pi_{\{D_1,D_2,\ldots,D_p\}}REL\_S$.

Then $REL\_R \bowtie REL\_S$
$= \{(t_i, s_i, t_j, \alpha_{i,1}(t) \cap \alpha_{j,1}(t), \alpha_{i,2}(t) \cap \alpha_{j,2}(t), \ldots,$
$\alpha_{i,k}(t) \cap \alpha_{j,k}(t)) \mid i \in I, j \in J, s_i = s_j\}$.

Informally, natural join computation can be thought of as a three-step process. First we compute the product. The second step involves selecting rows with identical values in their common relational attribute(s). Since common attributes are repeated, we delete repeating attributes in the third step. We now illustrate the join operation.

Let REL_R be a ciset relation on $R = \{A, B, C, F\}$ and REL_T be a ciset relation on $T = \{D, C, E, F\}$.

REL_R

A	B	C	F
a_1	b_1	c_1	$\langle 0.5, 0.8\rangle$
a_2	b_1	c_2	$\langle 0, 0.8\rangle$
a_3	b_2	c_1	$\langle 0.7, 0\rangle$
a_4	b_1	c_3	$\langle 1, 1\rangle$

REL_T

D	C	E	F
d_1	c_1	e_2	$\langle 0.4, 0.9\rangle$
d_2		e_3	$\langle 0.6, 0\rangle$
d_3	c_2	e_2	$\langle 0, 0.7\rangle$
d_4	c_2	e_1	$\langle 1, 1\rangle$

Step 1: Compute the product

In this case, there is only one common attribute, C. In what follows, we use $REL\_R.C$ to denote the attribute C in R and use $REL\_T.C$ to denote the attribute C in T respectively. In this step, we compute the product of REL_R and REL_T.

$REEL\_R \times REL\_T$

A	B	$REL\_R.C$	D	$REL\_T.C$	E	F
a_1	b_1	c_1	d_1	c_1	e_2	$\langle 0.5, 0.8\rangle$
a_2	b_1	c_2	d_1	c_1	e_2	$\langle 0.4, 0.8\rangle$
a_3	b_2	c_1	d_1	c_1	e_2	$\langle 0.7, 0\rangle$
a_4	b_1	c_3	d_1	c_1	e_2	$\langle 1, 0.9\rangle$
a_1	b_1	c_1	d_2		e_3	$\langle 0.6, 0\rangle$
a_2	b_1	c_2	d_2		e_3	$\langle 0.6, 0\rangle$
a_3	b_2	c_1	d_2		e_3	$\langle 0.7, 0\rangle$
a_1	b_1	c_1	d_3	c_2	e_2	$\langle 0.5, 0.7\rangle$
a_2	b_1	c_2	d_3	c_2	e_2	$\langle 0, 0.7\rangle$
a_3	b_2	c_1	d_3	c_2	e_2	$\langle 0.7, 0\rangle$
a_4	b_1	c_3	d_3	c_2	e_2	$\langle 1, 0.7\rangle$
a_1	b_1	c_1	d_4	c_2	e_1	$\langle 1, 0.8\rangle$
a_2	b_1	c_2	d_4	c_2	e_1	$\langle 1, 0.8\rangle$
a_4	b_1	c_3	d_4	c_2	e_1	$\langle 1, 1\rangle$

Step 2: Select rows with identical values in their common attribute(s)

$\sigma_{REL\_R.C=REL\_T.C} REL\_R \times REL\_T$

A	B	$REL\_R.C$	D	$REL\_T.C$	E	F
a_1	b_1	c_1	d_1	c_1	e_2	$\langle 0.5, 0.8\rangle$
a_3	b_2	c_1	d_1	c_1	e_2	$\langle 0.7, 0\rangle$
a_2	b_1	c_2	d_3	c_2	e_2	$\langle 0, 0.7\rangle$
a_2	b_1	c_2	d_4	c_2	e_1	$\langle 1, 0.8\rangle$

Step 3: Project on distinct attribute(s)

Attributes $REL\_R.C$ and $REL\_T.C$ are identical. Therefore, we eliminate the repeating attribute, say $REL\_T.C$. Further, we rename $REL\_R.C$ as C. Thus

$REL\_R \bowtie REL\_T$

A	B	C	D	E	F
a_1	b_1	c_1	d_1	e_2	$\langle 0.5, 0.8 \rangle$
a_3	b_2	c_1	d_1	e_2	$\langle 0.7, 0 \rangle$
a_2	b_1	c_2	d_3	e_2	$\langle 0, 0.7 \rangle$
a_2	b_1	c_2	d_4	e_1	$\langle 1, 0.8 \rangle$

Let REL_R be a relation on $R = \{A, B, C\}$ and REL_T be a relation on $T = \{D, C, E\}$ as shown below.

REL_R

A	B	C
a_1	b_1	c_1
a_2	b_1	c_2
a_3	b_2	c_1
a_4	b_1	c_3

REL_T

D	C	E
d_1	c_1	e_2
d_2		e_3
d_3	c_2	e_2
d_4	c_2	e_1

Then

F_{REL_R}

A	B	C	CI
a_1	b_1	c_1	$\langle 0, 1 \rangle$
a_2	b_1	c_2	$\langle 0, 1 \rangle$
a_3	b_2	c_1	$\langle 0, 1 \rangle$
a_4	b_1	c_3	$\langle 0, 1 \rangle$

F_{REL_T}

D	C	E	CI
d_1	c_1	e_2	$\langle 0, 1 \rangle$
d_2		e_3	$\langle 0, 1 \rangle$
d_3	c_2	e_2	$\langle 0, 1 \rangle$
d_4	c_2	e_1	$\langle 0, 1 \rangle$

Step 1: Compute the product

$F_{REL\_R} \times F_{REL\_T}$

A	B	$REL\_R.C$	D	$REL\_T.C$	E	CI
a_1	b_1	c_1	d_1	c_1	e_2	$\langle 0, 1 \rangle$
a_2	b_1	c_2	d_1	c_1	e_2	$\langle 0, 1 \rangle$
a_3	b_2	c_1	d_1	c_1	e_2	$\langle 0, 1 \rangle$
a_4	b_1	c_3	d_1	c_1	e_2	$\langle 0, 1 \rangle$
a_1	b_1	c_1	d_2		e_3	$\langle 0, 1 \rangle$
a_2	b_1	c_2	d_2		e_3	$\langle 0, 1 \rangle$
a_3	b_2	c_1	d_2		e_3	$\langle 0, 1 \rangle$
a_4	b_1	c_3	d_2		e_3	$\langle 0, 1 \rangle$
a_1	b_1	c_1	d_3	c_2	e_2	$\langle 0, 1 \rangle$
a_2	b_1	c_2	d_3	c_2	e_2	$\langle 0, 1 \rangle$
a_3	b_2	c_1	d_3	c_2	e_2	$\langle 0, 1 \rangle$
a_4	b_1	c_3	d_3	c_2	e_2	$\langle 0, 1 \rangle$
a_1	b_1	c_1	d_4	c_2	e_1	$\langle 0, 1 \rangle$
a_2	b_1	c_2	d_4	c_2	e_1	$\langle 0, 1 \rangle$
a_3	b_2	c_1	d_4	c_2	e_1	$\langle 0, 1 \rangle$
a_4	b_1	c_3	d_4	c_2	e_1	$\langle 0, 1 \rangle$

Step 2: Select rows with identical values in their common attribute(s)

$\sigma_{REL\_R.C=REL\_T.C} F_{REL\_R} \times F_{REL\_T}$

A	B	$REL\_R.C$	D	$REL\_T.C$	E	CI
a_1	b_1	c_1	d_1	c_1	e_2	$\langle 0,1\rangle$
a_3	b_2	c_1	d_1	c_1	e_2	$\langle 0,1\rangle$
a_2	b_1	c_2	d_3	c_2	e_2	$\langle 0,1\rangle$
a_2	b_1	c_2	d_4	c_2	e_1	$\langle 0,1\rangle$

Step 3: Project on distinct attribute(s)

Attributes $REL\_R.C$ and $REL\_T.C$ are identical. Therefore, we eliminate the repeating attribute, say $REL\_T.C$. Further, we rename $REL\_R.C$ as C. Thus

$F_{REL\_R} \bowtie F_{REL\_T}$

A	B	C	D	E	CI
a_1	b_1	c_1	d_1	e_2	$\langle 0,1\rangle$
a_3	b_2	c_1	d_1	e_2	$\langle 0,1\rangle$
a_2	b_1	c_2	d_3	e_2	$\langle 0,1\rangle$
a_2	b_1	c_2	d_4	e_1	$\langle 0,1\rangle$

Therefore,

$\left(F_{REL\_R} \bowtie F_{REL\_T}\right)^1$

A	B	C	D	E
a_1	b_1	c_1	d_1	e_2
a_3	b_2	c_1	d_1	e_2
a_2	b_1	c_2	d_3	e_2
a_2	b_1	c_2	d_4	e_1

and thus
$REL\_R \bowtie REL\_S = \left(F_{REL\_R} \bowtie F_{REL\_S}\right)^1.$

Theorem 4.3.2 *Let REL_R, REL_S two ciset relations on ciset relational schemes R and S respectively. Then the following diagram commute.*

$$\begin{array}{ccc} (REL\_R, REL\_S) & \xrightarrow{F} & (F_{REL\_R}, F_{REL\_S}) \\ \downarrow \bowtie & & \downarrow \bowtie \\ REL\_R \bowtie REL\_S & \xleftarrow{(\,)^1} & F_{REL\_R} \bowtie F_{REL\_S} \end{array}$$

Proof. Since natural join can be thought of a product operation followed by a selection and a project operation, the result follows. ■

4.4 Equi-join and Theta-join

Both *equi-join* and *theta-join* are binary operations. Let REL_R and REL_T be two ciset relations on ciset relational schemes $R = \{A_1, A_2, \ldots, A_n, C_1, \ldots, C_k\}$ and $T = \{B_1, B_2, \ldots, B_m, C_1, \ldots, C_k\}$ respectively where $\{A_1, A_2, \ldots, A_n\}$ and $\{B_1, B_2, \ldots, B_m\}$ are sets of ciset relational attributes; $\{C_1, C_2, \ldots, C_k\}$ is the set of attributes common to R and T such that $DOM(C_i) = \mathfrak{C}$.equi-join links tables REL_R and REL_T on the basis of an equality condition that compares specified attributes and produces a ciset relation on $\{A_1, A_2, \ldots, A_n, B_1, B_2, \ldots, B_m, C_1, \ldots, C_k\}$. The attributes involved in comparison must have identical domains. We use the symbol $[attribute_1 = attribute_2]$ for the equi-join operation. Unlike join operation, duplicate attributes are kept in this case. If the comparison operation is anything other than equality, it is called a theta-join.

Informally, equi-join (theta-join) computation can be thought of as a two steps process. First we compute the product. The second step involves selecting rows satisfying the condition specified. We now proceed to illustrate the equi-join operation.

Let REL_R be a ciset relation on $R = \{A, B, C1, F\}$ and REL_T be a ciset relation on $T = \{D, C2, E, F\}$, where $DOM(C1) = DOM(C2)$. Let us illustrate the computation of $REL\_R[C1 = C2]REL\_T$.

REL_R

A	B	$C1$	F
a_1	b_1	c_1	$\langle 0.5, 0.8\rangle$
a_2	b_1	c_2	$\langle 0, 0.8\rangle$
a_3	b_2	c_1	$\langle 0.7, 0\rangle$
a_4	b_1	c_3	$\langle 1, 1\rangle$

REL_T

D	$C2$	E	F
d_1	c_1	e_2	$\langle 0.4, 0.9\rangle$
d_2		e_3	$\langle 0.6, 0\rangle$
d_3	c_2	e_2	$\langle 0, 0.7\rangle$
d_4	c_2	e_1	$\langle 1, 1\rangle$

Step 1: Compute the product

$REL\_R \times REL\_T$

A	B	$C1$	D	$C2$	E	F
a_1	b_1	c_1	d_1	c_1	e_2	$\langle 0.5, 0.8\rangle$
a_2	b_1	c_2	d_1	c_1	e_2	$\langle 0.4, 0.8\rangle$
a_3	b_2	c_1	d_1	c_1	e_2	$\langle 0.7, 0\rangle$
a_4	b_1	c_3	d_1	c_1	e_2	$\langle 1, 0.9\rangle$
a_1	b_1	c_1	d_2		e_3	$\langle 0.6, 0\rangle$
a_2	b_1	c_2	d_2		e_3	$\langle 0.6, 0\rangle$
a_3	b_2	c_1	d_2		e_3	$\langle 0.7, 0\rangle$
a_1	b_1	c_1	d_3	c_2	e_2	$\langle 0.5, 0.7\rangle$
a_2	b_1	c_2	d_3	c_2	e_2	$\langle 0, 0.7\rangle$
a_3	b_2	c_1	d_3	c_2	e_2	$\langle 0.7, 0\rangle$
a_4	b_1	c_3	d_3	c_2	e_2	$\langle 1, 0.7\rangle$
a_1	b_1	c_1	d_4	c_2	e_1	$\langle 1, 0.8\rangle$
a_2	b_1	c_2	d_4	c_2	e_1	$\langle 1, 0.8\rangle$
a_4	b_1	c_3	d_4	c_2	e_1	$\langle 1, 1\rangle$

Step 2: Select rows with identical values in the indicated attribute(s)

$REL\_R[C1 = C2]REL\_T$

A	B	$C1$	D	$C2$	E	F
a_1	b_1	c_1	d_1	c_1	e_2	$\langle 0.5, 0.8\rangle$
a_3	b_2	c_1	d_1	c_1	e_2	$\langle 0.7, 0\rangle$
a_2	b_1	c_2	d_3	c_2	e_2	$\langle 0, 0.7\rangle$
a_2	b_1	c_2	d_4	c_2	e_1	$\langle 1, 0.8\rangle$

Let REL_R be a relation on $R = \{A, B, C\}$ and REL_T be a relation on $T = \{D, C, E\}$ as shown below.

REL_R

A	B	$C1$
a_1	b_1	c_1
a_2	b_1	c_2
a_3	b_2	c_1
a_4	b_1	c_3

REL_T

D	$C2$	E
d_1	c_1	e_2
d_2		e_3
d_3	c_2	e_2
d_4	c_2	e_1

Then

F_{REL_R}

A	B	$C1$	CI
a_1	b_1	c_1	$\langle 0, 1\rangle$
a_2	b_1	c_2	$\langle 0, 1\rangle$
a_3	b_2	c_1	$\langle 0, 1\rangle$
a_4	b_1	c_3	$\langle 0, 1\rangle$

F_{REL_T}

D	$C2$	E	CI
d_1	c_1	e_2	$\langle 0, 1\rangle$
d_2		e_3	$\langle 0, 1\rangle$
d_3	c_2	e_2	$\langle 0, 1\rangle$
d_4	c_2	e_1	$\langle 0, 1\rangle$

Step 1: Compute the product

$F_{REL\_R} \times F_{REL\_T}$

A	B	$C1$	D	$C2$	E	CI
a_1	b_1	c_1	d_1	c_1	e_2	$\langle 0,1\rangle$
a_2	b_1	c_2	d_1	c_1	e_2	$\langle 0,1\rangle$
a_3	b_2	c_1	d_1	c_1	e_2	$\langle 0,1\rangle$
a_4	b_1	c_3	d_1	c_1	e_2	$\langle 0,1\rangle$
a_1	b_1	c_1	d_2		e_3	$\langle 0,1\rangle$
a_2	b_1	c_2	d_2		e_3	$\langle 0,1\rangle$
a_3	b_2	c_1	d_2		e_3	$\langle 0,1\rangle$
a_4	b_1	c_3	d_2		e_3	$\langle 0,1\rangle$
a_1	b_1	c_1	d_3	c_2	e_2	$\langle 0,1\rangle$
a_2	b_1	c_2	d_3	c_2	e_2	$\langle 0,1\rangle$
a_3	b_2	c_1	d_3	c_2	e_2	$\langle 0,1\rangle$
a_4	b_1	c_3	d_3	c_2	e_2	$\langle 0,1\rangle$
a_1	b_1	c_1	d_4	c_2	e_1	$\langle 0,1\rangle$
a_2	b_1	c_2	d_4	c_2	e_1	$\langle 0,1\rangle$
a_3	b_2	c_1	d_4	c_2	e_1	$\langle 0,1\rangle$
a_4	b_1	c_3	d_4	c_2	e_1	$\langle 0,1\rangle$

Step 2: Select rows with identical values in the attributes specified

$\sigma_{C1=C2} F_{REL\_R} \times F_{REL\_T}$

A	B	$C1$	D	$C2$	E	CI
a_1	b_1	c_1	d_1	c_1	e_2	$\langle 0,1\rangle$
a_3	b_2	c_1	d_1	c_1	e_2	$\langle 0,1\rangle$
a_2	b_1	c_2	d_3	c_2	e_2	$\langle 0,1\rangle$
a_2	b_1	c_2	d_4	c_2	e_1	$\langle 0,1\rangle$

Therefore,

$\left(F_{REL\_R}[C1 = C2]F_{REL\_T}\right)^1$

A	B	$C1$	D	$C2$	E
a_1	b_1	c_1	d_1	c_1	e_2
a_3	b_2	c_1	d_1	c_1	e_2
a_2	b_1	c_2	d_3	c_2	e_2
a_2	b_1	c_2	d_4	c_2	e_1

and thus

$REL\_R[C1 = C2]REL\_S = \left(F_{REL\_R}[C1 = C2]F_{REL\_S}\right)^1.$

Theorem 4.4.1 *Let REL_R, REL_S two ciset relations on ciset relational schemes R and S respectively. Then the following diagram commute.*

$$\begin{array}{ccc} (REL\_R, REL\_S) & \xrightarrow{F} & (F_{REL\_R}, F_{REL\_S}) \\ [\theta - join\ constraints] \downarrow & & \downarrow [\theta - join\ constraints] \\ REL\_R \bowtie REL\_S & \xleftarrow{(\,)^1} & F_{REL\_R} \bowtie F_{REL\_S} \end{array}$$

Proof. Since equi-join can be thought of a product operation followed by a selection, the result follows. ■

4.5 Divide

Divide operation is very similar to integer division. Consider the integer division of 17 by 5. The result in this case is 3. There are two important points worth consider in this regard.

- $17 \geq 5 \times 3$;
- $17 < 5 \times n$, for all integers $n > 3$.

Thus 3 is the largest integer value of n such that $17 \geq 5 \times n$.

Divide is a binary operation and we will use the notation $\div$. Let REL_R and REL_T be two ciset relations on ciset relational schemes
$R = \{A_1, A_2, \ldots, A_n, B_1, B_2, \ldots, B_m, C_1, C_2, \ldots, C_k\}$
and
$T = \{B_1, B_2, \ldots, B_m, C_1, C_2, \ldots, C_k\}$
respectively, where $\{C_1, C_2, \ldots, C_k\}$ is the set of attributes common to R and T such that $DOM(C_i) = \mathfrak{C}$. Then $REL\_S = REL\_R \div REL\_T$ is defined and is a ciset relation on the ciset relational scheme $S = \{A_1, A_2, \ldots, A_n, C_1, C_2, \ldots, C_k\}$ such that the following conditions hold.

- $REL\_R \supseteq REL\_T \times REL\_S$;
- $REL\_R \subset REL\_T \times REL\_Q$, for all ciset relations $REL\_Q \supset REL\_S$.

Informally, divide computation can be thought of as a three-step process. First for each tuple t of REL_T, select all tuples r of REL_R such $r(T) = t$ and $t(C_1) \preceq r(C_i)$, for $i = 1, 2, \ldots, k$. Recall that T is the ciset relational scheme of REL_T. Let us denote the ciset relation so obtained by REL_R_t. The second step involves projecting all ciset relations REL_R_t on attributes of S. Thus by the end of step 2, for each tuple t of REL_T, we have a ciset relation $\Pi_S REL\_R_t$ on the ciset relational scheme S. The final step is the intersection of all ciset relations $\Pi_S REL\_R_t$.

Let REL_R be a ciset relation on $R = \{A, B, C\}$ and REL_T be a ciset relation on $T = \{A, C\}$.

REL_R

A	B	C
a_1	b_1	$\langle 0.4, 0.8\rangle$
a_2	b_1	$\langle 0.4, 0.9\rangle$
a_3	b_2	$\langle 0.5, 0.8\rangle$
a_4	b_3	$\langle 0.3, 0.6\rangle$
a_4	b_8	$\langle 0.2, 0.5\rangle$
a_5	b_8	$\langle 0.5, 0.3\rangle$
a_2	b_2	$\langle 0.7, 0.4\rangle$
a_1	b_4	$\langle 0.9, 0.5\rangle$
a_1	b_2	$\langle 0.6, 0.7\rangle$

REL_T

A	C
a_1	$\langle 0.7, 0.6\rangle$
a_2	$\langle 0.8, 0.4\rangle$

Step 1: For each tuple t of REL_T, select all tuples r of REL_R such $r(T) = t$ and $t(C) \preceq r(C)$.

Let $t = a_1$. Select tuples of REL_R such that attribute A is $a_1$and C value $\succeq \langle 0.7, 0.6\rangle$. That is, $\sigma_{(A=a_1)\wedge(C\succeq\langle 0.7,0.6\rangle)}REL\_R$.

$\sigma_{(A=a_1)\wedge(C\succeq\langle 0.7,0.6\rangle)}REL\_R$

A	B	C
a_1	b_1	$\langle 0.4, 0.8\rangle$
a_1	b_2	$\langle 0.6, 0.7\rangle$

Now choose t as a_2. Select tuples of REL_R such that attribute A is $a_2$and C value $\succeq \langle 0.8, 0.4\rangle$. That is, $\sigma_{(A=a_2)\wedge(C\succeq\langle 0.8,0.4\rangle)}REL\_R$.

$_{(A=a_2)\wedge(C\succeq\langle 0.8,0.4\rangle)}REL\_R$

A	B	C
a_2	b_1	$\langle 0.4, 0.9\rangle$
a_2	b_2	$\langle 0.7, 0.4\rangle$

Step 2: Project all ciset relations obtained in step 1 on attributes of S.

In this case, $S = \{B, C\}$, the set of attributes that are in REL_R but not in REL_T. Thus we have the following.

$\Pi_{\{B,C\}}\sigma_{(A=a_1)\wedge(C\succeq\langle 0.7,0.6\rangle)}REL\_R$

B	C
b_1	$\langle 0.4, 0.8\rangle$
b_2	$\langle 0.6, 0.7\rangle$

and

$\Pi_{\{B,C\}}\sigma_{(A=a_2)\wedge(C\succeq\langle 0.8,0.4\rangle)}REL\_R$

B	C
b_1	$\langle 0.4, 0.9\rangle$
b_2	$\langle 0.7, 0.4\rangle$

Step 3: Compute the intersection of all ciset relations obtained in step 2.

It is quite clear that intersection of $\Pi_{\{B,C\}}\sigma_{(A=a_1)\wedge(C\succeq\langle 0.7,0.6\rangle)}REL\_R$ and $\Pi_{\{B,C\}}\sigma_{(A=a_2)\wedge(C\succeq\langle 0.8,0.4\rangle)}REL\_R$ is the following:

$REL\_R \div REL\_T$

B	C
b_1	$\langle 0.4, 0.8\rangle$
b_2	$\langle 0.7, 0.4\rangle$

Let REL_R be a relation on $R = \{A, B\}$ and REL_T be a relation on $T = \{A\}$ as shown below.

REL_R

A	B
a_1	b_1
a_2	b_1
a_3	b_2
a_4	b_3
a_4	b_8
a_5	b_8
a_2	b_2
a_1	b_4
a_1	b_2

REL_T

A
a_1
a_2

F_{REL_R}

A	B	CI
a_1	b_1	$\langle 0, 1\rangle$
a_2	b_1	$\langle 0, 1\rangle$
a_3	b_2	$\langle 0, 1\rangle$
a_4	b_3	$\langle 0, 1\rangle$
a_4	b_8	$\langle 0, 1\rangle$
a_5	b_8	$\langle 0, 1\rangle$
a_2	b_2	$\langle 0, 1\rangle$
a_1	b_4	$\langle 0, 1\rangle$
a_1	b_2	$\langle 0, 1\rangle$

F_{REL_T}

A	CI
a_1	$\langle 0, 1\rangle$
a_2	$\langle 0, 1\rangle$

Step 1: For each tuple t of F_{REL_T}, select all tuples r of F_{REL_R} such $r(T) = t$ and $t(CI) \preceq r(CI)$.

Let $t = a_1$. Select tuples of F_{REL_R} such that attribute A is $a_1$and CI value $\succeq \langle 0, 1\rangle$. That is, $\sigma_{(A=a_1)\wedge(CI\succeq\langle 0,1\rangle)}F_{REL\_R}$.

$\sigma_{(A=a_1)\wedge(CI\succeq\langle 0,1\rangle)}F_{REL\_R}$

A	B	CI
a_1	b_1	$\langle 0,1\rangle$
a_1	b_4	$\langle 0,1\rangle$
a_1	b_2	$\langle 0,1\rangle$

Now choose t as a_2. Select tuples of F_{REL_R} such that attribute A is $a_2$and CI value $\succeq \langle 0,1\rangle$. That is, $\sigma_{(A=a_2)\wedge(CI\succeq\langle 0,1\rangle)}F_{REL\_R}$.

$_{(A=a_2)\wedge(CI\succeq\langle 0,1\rangle)}F_{REL\_R}$

A	B	CI
a_2	b_1	$\langle 0,1\rangle$
a_2	b_2	$\langle 0,1\rangle$

Step 2: Project all ciset relations obtained in step 1 on attributes of F_{REL_S}.

In this case, we have $\{B, CI\}$. Thus we have the following.

$\Pi_{B,CI}\sigma_{(A=a_1)\wedge(CI\succeq\langle 0,1\rangle)}F_{REL\_R}$

B	CI
b_1	$\langle 0,1\rangle$
b_4	$\langle 0,1\rangle$
b_2	$\langle 0,1\rangle$

and

$\Pi_{B,CI}\sigma_{(A=a_2)\wedge(CI\succeq\langle 0,1\rangle)}F_{REL\_R}$

B	CI
b_1	$\langle 0,1\rangle$
b_2	$\langle 0,1\rangle$

Step 3: Compute the intersection of all ciset relations obtained in step 2.

It is quite clear that intersection of $\Pi_{B,CI}\sigma_{(A=a_1)\wedge(CI\succeq\langle 0,1\rangle)}F_{REL_R}$and $\Pi_{B,CI}\sigma_{(A=a_2)\wedge(CI\succeq\langle 0,1\rangle)}F_{REL\_R}$ is the following:

$F_{REL\_R} \div F_{REL\_T}$

B	CI
b_1	$\langle 0,1\rangle$
b_2	$\langle 0,1\rangle$

Therefore,

$\left(F_{REL\_R} \div F_{REL\_T}\right)^1$

B
b_1
b_2

and thus
$REL\_R \div REL\_S = \left(F_{REL\_R} \div F_{REL\_S}\right)^1$ in this example.

Theorem 4.5.1 *Let REL_R, REL_S any two ciset relations on ciset relational schemes R and S respectively. Then the following diagram commute.*

$$\begin{array}{ccc} (REL\_R, REL\_S) & \xrightarrow{F} & (F_{REL\_R}, F_{REL\_S}) \\ \downarrow \div & & \downarrow \div \\ REL\_R \div REL\_S & \xleftarrow{(\ )^1} & F_{REL\_R} \div F_{REL\_S} \end{array}$$

Proof. Since divide operation can be thought of a sequence of selections followed by an intersection, the result follows. ■

The following result holds and proof follows from our above discussion. Hence we state the result without proof. In short, the result shows that starting from a set of relations, you may either convert them to ciset relations and apply the corresponding ciset relational operation or you may first perform the relational operation and then convert it to a ciset relation. This result is important from a practical point of view. If part of a query involves only relations, the partial computation can be carried out using relational operations and then converted to a ciset relation and integrated with the other partial results of the query computation.

Theorem 4.5.2 *Let REL_R, REL_S any two relations on a relational scheme R. Then the following diagrams commute.*

$$\begin{array}{ccc} (REL\_R, REL\_S) & \xrightarrow{F} & (F_{REL\_R}, F_{REL\_S}) \\ \downarrow \cup & & \downarrow \cup \\ REL\_R \cup REL\_S & \xrightarrow{F} & F_{REL\_R} \cup F_{REL\_S} \end{array}$$

$$\begin{array}{ccc} (REL\_R, REL\_S) & \xrightarrow{F} & (F_{REL\_R}, F_{REL\_S}) \\ \downarrow \cap & & \downarrow \cap \\ REL\_R \cap REL\_S & \xrightarrow{F} & F_{REL\_R} \cap F_{REL\_S} \end{array}$$

$$\begin{array}{ccc} (REL\_R, REL\_S) & \xrightarrow{F} & (F_{REL\_R}, F_{REL\_S}) \\ \downarrow - & & \downarrow - \\ REL\_R - REL\_S & \xrightarrow{F} & F_{REL\_R} - F_{REL\_S} \end{array}$$

$$\begin{array}{ccc} REL\_R & \xrightarrow{F} & F_{REL\_R} \\ \downarrow{\sigma} & & \downarrow{\sigma} \\ \sigma_{criteria}REL\_R & \xrightarrow{F} & \sigma_{criteria}F_{REL\_R} \end{array}$$

$$\begin{array}{ccc} REL\_R & \xrightarrow{F} & F_{REL\_R} \\ \downarrow{\Pi} & & \downarrow{\Pi} \\ \Pi_{attributes}REL\_R & \xrightarrow{F} & \Pi_{attributes}F_{REL\_R} \end{array}$$

$$\begin{array}{ccc} (REL\_R, REL\_S) & \xrightarrow{F} & (F_{REL\_R}, F_{REL\_S}) \\ \downarrow{\times} & & \downarrow{\times} \\ REL\_R \times REL\_S & \xrightarrow{F} & F_{REL\_R} \times F_{REL\_S} \end{array}$$

$$\begin{array}{ccc} (REL\_R, REL\_S) & \xrightarrow{F} & (F_{REL\_R}, F_{REL\_S}) \\ \downarrow{\bowtie} & & \downarrow{\bowtie} \\ REL\_R \bowtie REL\_S & \xrightarrow{F} & F_{REL\_R} \bowtie F_{REL\_S} \end{array}$$

$$\begin{array}{ccc} (REL\_R, REL\_S) & \xrightarrow{F} & (F_{REL\_R}, F_{REL\_S}) \\ [\theta - join\ constraints]\downarrow & & \downarrow[\theta - join\ constraints] \\ REL\_R \bowtie REL\_S & \xrightarrow{F} & F_{REL\_R} \bowtie F_{REL\_S} \end{array}$$

$$\begin{array}{ccc} (REL\_R, REL\_S) & \xrightarrow{F} & (F_{REL\_R}, F_{REL\_S}) \\ \downarrow{\div} & & \downarrow{\div} \\ REL\_R \div REL\_S & \xrightarrow{F} & F_{REL\_R} \div F_{REL\_S} \end{array}$$

5
ALTERNATE WORLDS

5.1 Introduction

In this Chapter we will formalize the semantics of our model. Closely related notions of

- representation,
- possibility functions,
- alternate worlds

have been the tools used by leading researchers to formalize the information content of a databases with incomplete information [1, 3, 7, 9, 11, 12, 14, 17, 22, 23, 24, 28, 35, 43, 44]. The approach we adopt in this instance is similar to theirs. To be specific, we use the notion of alternate worlds to formalize the information content of a ciset relational database. A ciset relation represents a set of (regular) relations. Once this set has been identified, a query on ciset relations can be as well processed against the set of (regular) relations represented by ciset relations involved. Clearly, this approach is computationally inefficient and is not an approach we would recommend. On the other hand, this approach will well explain the semantics of ciset relational model in a formal setting. From a practical point of view, what we need is a query processing methodology that can be applied directly to ciset relations. Further, we would like to generate the same answers that would be obtained by processing the query against represented (regular) relations. A query processing methodology that satisfies the above condition is called precise.

In what follows, we will characterize the set of (regular) relations represented by a ciset relation.

Given a ciset what are the alternate worlds? We will answer this question first, before we tackle the problem at hand. We begin our discussion with the following example.

Example 5.1.1 *Let $S = \{u, v, x, y, z\}$ and let F be a ciset on S defined by* $F(u) = \langle 0.7, 0.3\rangle, F(v) = \langle 0, 1\rangle, F(x) = \langle 0, 0\rangle, F(y) = \langle 0.4, 0.8\rangle, F(z) = \langle 1, 1\rangle$.

Then

$$F_s^t = \begin{cases} \{v, x\} & 0 < s \leq 0.4; t = 0 \\ \{v\} & 0 < s \leq 0.4; 0 < t \leq 0.3 \\ \{v\} & 0 < s \leq 0.4; 0.3 < t \leq 0.8 \\ \{v\} & 0 < s \leq 0.4; 0.8 < t \leq 1 \\ \{v, x, y\} & 0.4 < s \leq 0.7; t = 0 \\ \{v, y\} & 0.4 < s \leq 0.7; 0 < t \leq 0.3 \\ \{v, y\} & 0.4 < s \leq 0.7; 0.3 < t \leq 0.8 \\ \{v\} & 0.4 < s \leq 0.7; 0.8 < t \leq 1 \\ \{u, v, x, y\} & 0.7 < s \leq 1; t = 0 \\ \{u, v, y\} & 0.7 < s \leq 1; 0 < t \leq 0.3 \\ \{v, y\} & 0.7 < s \leq 1; 0.3 < t \leq 0.8 \\ \{v\} & 0.7 < s \leq 1; 0.8 < t \leq 1 \\ S & 1 < s \leq 1 + \varepsilon; t = 0 \\ \{u, v, y, z\} & 1 < s \leq 1 + \varepsilon; 0 < t \leq 0.3 \\ \{v, y, z\} & 1 < s \leq 1 + \varepsilon; 0.3 < t \leq 0.8 \\ \{v, z\} & 1 < s \leq 1 + \varepsilon; 0.8 < t \leq 1 \end{cases}$$

Thus a ciset F on S can be considered as a collection of subsets of S.

FIGURE 5.1 Graph of F.

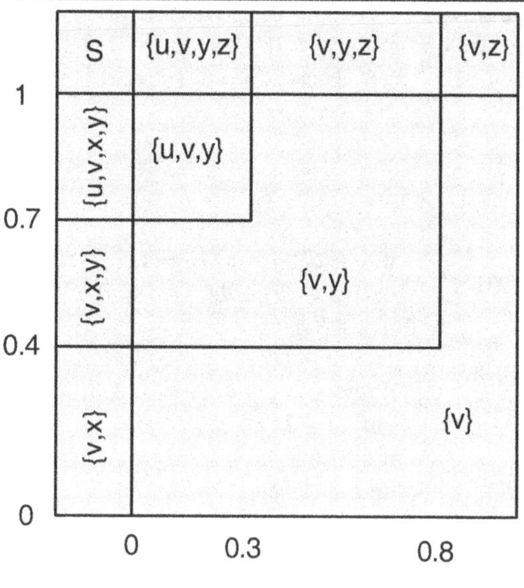

The following fact is worth noticing. The ciset F is completely determined by two descending chains of subsets of S, starting with S. In the above example, the descending chains are

$S \supset \{u, v, x, y\} \supset \{v, x, y\} \supset \{v, x\}$ *and* $S \supset \{u, v, y, z\} \supset \{v, y, z\} \supset \{v, z\}$.

Also note that $F_s^t = F^t \cap F_s$, *for all* $0 < s \leq 1 + \varepsilon, 0 \leq t \leq 1$.

Proposition 5.1.2 *Let F be a ciset on a nonempty set S. Then for $0 < s < 1+\varepsilon, 0 \leq t \leq 1, F_s^t = F^t \cap F_s$. Further, if the range of F is finite, there exists $0 \leq s_0 < \ldots < s_p \leq 1 < s_{p+1} = 1 + \varepsilon$ and $0 \leq t_0 < \ldots < t_q \leq 1$ such that $\varnothing = F_{s_0} \subset F_{s_1} \subset \ldots \subset F_{s_p} \subset F_{s_{p+1}} = S$ and $S = F^{t_0} = F^0 \supset F^{t_1} \supset F^{t_2} \supset \ldots \supset F^{t_q} = F^1 \supseteq \varnothing$.*

Example 5.1.3 *Consider the ciset F given in Figure 5.1. Observe that $s_0 = 0 < s_1 = 0.4 < s_2 = 0.7 < s_3 = 1 < s_4 = 1 + \varepsilon$ and $F_{s_0} = F_0 = \varnothing \subset F_{s_1} = F_{0.4} = \{v, x\} \subset F_{s_2} = F_{0.7} = \{v, x, y\} \subset F_{s_3} = F_1 = \{u, v, x, y\} \subset F_{s_4} = F_{1+\varepsilon} = S$. Thus $p = 3$. Similarly, $t_0 = 0 < t_1 = 0.3 < t_2 = 0.8 < t_3 = 1$ and $F^{t_0} = F^0 = S \supseteq F^{t_1} = F^{0.3} = \{u, v, y, z\} \supset F^{t_2} = F^{0.8} = \{v, y, z\} \supset F^{t_3} = F^1 = \{v, z\} \supseteq \varnothing$. Thus $q = 3$.*

From the above proposition, we see that a finite-valued ciset F on S determines two chains of subsets of S and $p+q+3$ real numbers: $0 \leq s_0 < \ldots < s_p \leq 1 < s_{p+1} = 1 + \varepsilon, 0 \leq t_0 < \ldots < t_q \leq 1$ Conversely, given two finite chains of subsets $C_0 = \varnothing \subset C_1 \subset \ldots \subset C_p \subset C_{p+1} = S, p > 0; S = D_0 \supset D_1 \supset \ldots \supset D_q \supseteq \varnothing, q > 0$; there exits a ciset F on S such that

$C_i, i = 1, 2, \ldots, p+1$ are the lower cut sets of F and $D_j, j = 0, 1, 2, \ldots, q$ are the upper cuts sets of F. The construction of F can be outlined as follows. Choose real numbers $s_i, i = 0, 1, \ldots, p, p+1$; $t_j, j = 0, 1, \ldots, q$ such that $0 \leq s_0 < s_1 < \ldots < s_p \leq 1 < s_{p+1} = 1 + \varepsilon$ and $0 \leq t_0 < \ldots < t_q \leq 1$. For each $w \in S$,

$$F(\omega) = \begin{cases} \langle s_i, t_j \rangle & \text{if } \omega \in C_{i+1} - C_i, i = 1, 2, 3, \ldots, p \\ & \text{and } \omega \in D_j - D_{j+1}, j = 0, 1, 2, \ldots, q-1; \\ \langle s_0, t_j \rangle & \text{if } \omega \in C_1 \\ & \text{and } \omega \in D_j - D_{j+1}, j = 0, 1, 2, \ldots, q-1; \\ \langle s_i, t_q \rangle & \text{if } \omega \in C_{i+1} - C_i, i = 1, 2, 3, \ldots, p \\ & \text{and } \omega \in D_q; \\ \langle s_0, t_q \rangle & \text{if } \omega \in C_1 \\ & \text{and } \omega \in D_q. \end{cases}$$

Thus it follows that a finite-valued ciset is completely determined by two chains of subsets of S and set of values $0 = s_0 < \ldots < s_p \leq 1 < s_{p+1} = 1+\varepsilon$ and $0 \leq t_0 < \ldots < t_q \leq 1$. Note that we set s_{p+1} always as 1 and thus C_{p+1} is always S for notational convenience only.

Example 5.1.4 *Let $p = q = 3$ and let $S = \{u, v, x, y, z\}$.Assume that $C_0 = \varnothing \subset C_1 = \{v, x\} \subset C_2 = \{v, x, y\} \subset C_3 = \{u, v, x, y\} \subset C_4 = S$ and $D_0 = S \supset D_1 = \{u, v, y, z\} \supset D_2 = \{v, y, z\} \supset D_3 = \{v, z\} \supseteq \varnothing$. Further, let $s_0 = 0 < s_1 = 0.4 < s_2 = 0.7 < s_3 = 1 < s_4 = 1 + \varepsilon$ and $t_0 = 0 < t_1 = 0.3 < t_2 = 0.8 < t_3 = 1$. Note that $u \in C_3 - C_2$ and $u \in D_1 - D_2$. Therefore $F(u) = \langle s_2, t_1 \rangle = \langle 0.7, 0.3 \rangle$. Similarly, $v \in C_1$ and $v \in D_3$. Therefore $F(v) = \langle s_0, t_q \rangle = \langle 0, 1 \rangle$. Now $x \in C_1$ and $x \in D_0 - D_1$. Therefore $F(u) = \langle s_0, t_0 \rangle = \langle 0, 0 \rangle$. Further $y \in C_2 - C_1$ and $y \in D_2 - D_3$. Therefore $F(y) = \langle s_1, t_2 \rangle = \langle 0.4, 0.8 \rangle$. Notice that $z \in C_4 - C_3$ and $z \in D_3$. Therefore $F(z) = \langle s_3, t_3 \rangle = \langle 1, 1 \rangle$.*

Example 5.1.5 *Let G be a ciset on S defined by*
$G(u) = \langle 0.3, 0.7 \rangle, G(v) = \langle 0.5, 0.6 \rangle, G(x) = \langle 0.2, 0.7 \rangle$,
$G(y) = \langle 0.2, 0.3 \rangle, G(z) = \langle 0.5, 0.6 \rangle$.
Then $G_1 = S, G_{0.5} = \{u, x, y\}, G_{0.3} = \{x, y\}$ and $G_{0.2} = \varnothing$. Similarly, $G^{0.3} = S, G^{0.6} = \{u, v, x, z\}, G^{0.7} = \{u, x\}$ and $G^1 = \varnothing$ and

$$G_s^t = \begin{cases} \varnothing & 0 < s \leq 0.2; 0 \leq t \leq 1 \\ \{x, y\} & 0.2 < s \leq 0.3; 0 \leq t \leq 0.3 \\ \{x\} & 0.2 < s \leq 0.3; 0.3 < t \leq 0.7 \\ \varnothing & 0.2 < s \leq 0.3; 0.7 < t \leq 1 \\ \{u, x, y\} & 0.3 < s \leq 0.5; 0 \leq t \leq 0.3 \\ \{u, x\} & 0.3 < s \leq 0.5; 0.3 < t \leq 0.7 \\ \varnothing & 0.3 < s \leq 0.5; 0.7 < t \leq 1 \\ S & 0.5 < s \leq 1; 0 \leq t \leq 0.3 \\ \{u, v, x, z\} & 0.5 < s \leq 1; 0.3 < t \leq 0.6 \\ \{u, x\} & 0.5 < s \leq 1; 0.6 < t \leq 0.7 \\ \varnothing & 0.5 < s \leq 1; 0.7 < t \leq 1 \end{cases}$$

The graph of G is given below.

FIGURE 5.2 Graph of G.

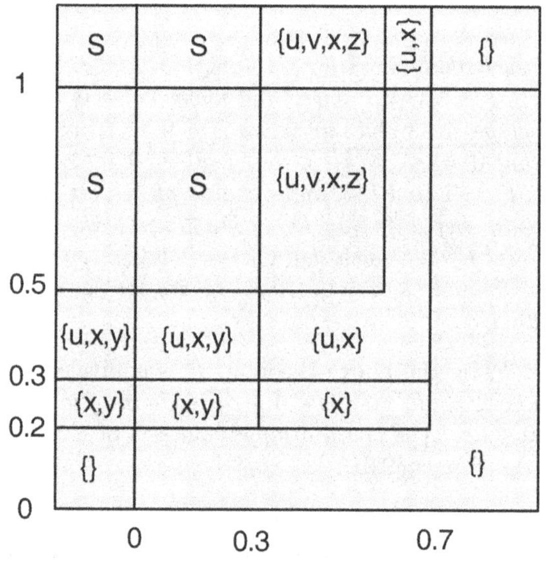

Conversely, given two chains of subsets $C_3 = S \supset C_2 = \{u, x, y\} \supset C_1 = \{x, y\} \supset C_0 = \varnothing, D_0 = S \supset D_1 = \{u, v, x, z\} \supset D_2 = \{u, x\} \supset D_3 = \varnothing$ *and real numbers* $0.2 < 0.3 < 0.5, 0.3 < 0.6 < 0.7$, *we notice that* $p = 2$ *and* $q = 2$. *We assign* $s_0 = 0.2, s_1 = 0.3, s_2 = 0.5, s_3 = 1 + \varepsilon$ *and* $s_4 = 1$. *Similarly,* $t_0 = 0.3, t_1 = 0.6, t_2 = 0.7$.

Note that $u \in C_2 - C_1$ *and* $u \in D_2 - D_3$. *Therefore* $F(u) = \langle s_1, t_2 \rangle = \langle 0.3, 0.7 \rangle$. *Similarly,* $v \in C_3 - C_2$ *and* $v \in D_1 - D_2$. *Therefore* $F(v) = \langle s_2, t_1 \rangle = \langle 0.5, 0.6 \rangle$. *Now* $x \in C_1$ *and* $x \in D_2 - D_3$. *Therefore* $F(u) = \langle s_0, t_2 \rangle = \langle 0.2, 0.7 \rangle$. *Further* $y \in C_1$ *and* $y \in D_0 - D_1$. *Therefore* $F(y) =$

$\langle s_0, t_0\rangle = \langle 0.2, 0.3\rangle$. *Notice that* $z \in C_3 - C_2$ *and* $z \in D_1 - D_2$. *Therefore* $F(z) = \langle s_2, t_1\rangle = \langle 0.5, 0.6\rangle$.

In our above discussion, we have the following facts. Let H be a ciset. Then

- $t_0 = \wedge\{u(H(x)) \mid x \in S\}$,
- $t_q = \vee\{u(H(x)) \mid x \in S\}$,
- $s_0 = \wedge\{l(H(x)) \mid x \in S\}$,
- $s_p = \vee\{l(H(x)) \mid x \in S\}$,
- s_{p+1} is defined as $1+\varepsilon$. As a consequence, $H_{s_{p+1}}$ is equal to S.

The following theorem is quite clear and hence we state it without any proof.

Theorem 5.1.6 *Let F be a finite-valued ciset on a nonempty set S. Then there exists $0 \leq s_0 < s_1 < \ldots < s_p \leq 1 < s_{p+1} = 1+\varepsilon$ and $0 \leq t_0 < \ldots < t_q \leq 1$ such that $S = F_{s_{p+1}} \supset F_{s_p} \supset \ldots \supset F_{s_1} \supset F_{s_0} = \varnothing$; $\varnothing \subseteq F^{t_q} \subset \ldots \subset F^{t_0} = S$ are lower and upper cut sets of F respectively. Conversely, given $p+q+3$ real numbers $s_i, i = 0, 1, \ldots, p$; $t_j, j = 0, 1, \ldots, q$ in $[0, 1]$ such that $0 = s_0 < s_1 < \ldots < s_p \leq 1 < s_{p+1} = 1+\varepsilon$ and $0 = t_0 < \ldots < t_q = 1$ and two finite chains of subsets $\varnothing = C_0 \subset C_1 \subset \ldots \subset C_p \subset C_{p+1} = S, p > 0; S = D_0 \supset D_1 \supset \ldots \supset D_q \supseteq \varnothing, q > 0$; there exits a unique ciset F on S such that $C_i, i = 0, 1, 2, \ldots, p+1$ are the set of lower cut sets of F and $D_j, j = 0, 1, 2, \ldots, q$ are the set of upper cuts sets of F respectively.*

Given a ciset F, by the above theorem, F is equivalent to an ordered pair $(\mathfrak{L}, \mathfrak{U})$, where $\mathfrak{L} = \{F_s \mid 0 < s < 1+\varepsilon\}$ is *chain of lower cut sets* of F and $\mathfrak{U} = \{F^t \mid 0 \leq t \leq 1\}$ is *chain of upper cut sets* of F respectively. Now define F° as ordered pair $(\mathfrak{L}, \mathfrak{U})$.

Definition 5.1.7 *A ciset F is said to represent F°. We indicate this as $rep(F) = F^\circ$. The set of subsets defined by $F^\circ(s,t) = F_s \cap F^t$, where $F_s \in \mathfrak{L}, F^t \in \mathfrak{U}, 0 < s < 1+\varepsilon$ and $0 \leq t \leq 1$ is called the* alternate worlds *of F.*

Note that in Example 5.1.1, if we define $F^\circ(s,t) = F_s^t$ for $0 < s < 1+\varepsilon$ and $0 \leq t \leq 1$, then $\{F^\circ(s,t) \mid 0 < s < 1+\varepsilon, 0 \leq t \leq 1\}$ is the alternate worlds of F. Further, the Figure 5.1 is a visual presentation of F°.

Example 5.1.8 *The ciset F presented in Example 5.1.1 in fact represents a collection of sets. They are*
$\{v\}, \{v, x\}, \{v, y\}, \{v, z\}, \{v, x, y\}, \{v, y, z\}$,
$\{u, v, y\}, \{u, v, x, y\}, \{u, v, y, z\}, S$.
Similarly, The ciset G presented in Example 5.1.5 represents the following collection of sets: $\varnothing, \{x\}, \{u, x\}, \{x, y\}, \{u, x, y\}, \{u, v, x, z\}, S$.

Having defined the notion of alternate worlds, we are now ready to introduce the concept of precision of ciset operations. In formally, an operation is precise if the extended version can be computed using the (regular) version of the operation.

Definition 5.1.9 *Let $\otimes$ be a unary set theory operation. A unary ciset relational operation $\otimes'$ is said to be precise, if*

$$rep(\otimes'(F)) = \otimes(rep(F))$$

for all ciset F (on S), where $\otimes(rep(F))$ represents a function f such that $f(s,t) = \otimes(F^\circ(s,t))$, $\forall 0 < s < 1+\varepsilon, 0 \leq t \leq 1$.

The above definition can be graphically depicted as follows:

$$\begin{array}{ccc} F & \xrightarrow{\otimes'} & \otimes'(F) \\ rep\big\downarrow & & \big\downarrow rep \\ rep(F) & \xrightarrow{\otimes} & rep(\otimes'(F)) = \otimes(rep(F)) \end{array}$$

In other words, let $\otimes$ be a unary operation on sets and let $\otimes'$ be the corresponding operation on cisets. Then given any ciset F, we can convert F to a collection of sets (which we call the alternate worlds) and apply $\otimes$ to obtain a collection of sets which is the same as the alternate worlds of $\otimes'(F)$.

Definition 5.1.10 *Let $\otimes$ be a binary set theory operation. A binary ciset theory operation $\otimes'$ is said to be precise, if*

$$rep(F \otimes' G) = rep(F) \otimes rep(G)$$

for all ciset F, G (on S), where $rep(F) \otimes rep(G)$ represents a function f such that $f(s,t) = F^\circ(s,t) \otimes G^\circ(s,t)$, $\forall 0 < s < 1+\varepsilon, 0 \leq t \leq 1$.

The above definition can be graphically depicted as follows:

$$\begin{array}{ccc} (F,G) & \xrightarrow{\otimes'} & F \otimes' G \\ rep\big\downarrow & & \big\downarrow rep \\ (rep(F), rep(G)) & \xrightarrow{\otimes} & rep(F \otimes' G) = rep(F) \otimes rep(G) \end{array}$$

In other words, let $\otimes$ be a binary operation on sets and let $\otimes'$ be the corresponding operation on cisets. Then given any two ciset F and G, we can convert F and G to two collection of sets (which we call the alternate worlds) and apply $\otimes$ to obtain a collection of sets which is the same as the alternate worlds of $F \otimes' G$.

5.2 Precision of Ciset Relational Operations

We now proceed to check whether or not the ciset operations introduced in Chapter 1 are precise.

Union

In this subsection, we show that the union operation is precise. We begin our discussion with the following motivating example.

Example 5.2.1 *Consider the ciset F and G presented in Example 5.1.1 and in Example 5.1.5 respectively. It is quite easy to observe that*
$(F \cup G)(u) = \langle 0.3, 0.7 \rangle$,
$(F \cup G)(v) = \langle 0, 1.0 \rangle$,
$(F \cup G)(x) = \langle 0, 0.7 \rangle$,
$(F \cup G)(y) = \langle 0.2, 0.8 \rangle$
and
$(F \cup G)(z) = \langle 0.5, 1.0 \rangle$.

FIGURE 5.3 Graph of $F \cup G$.

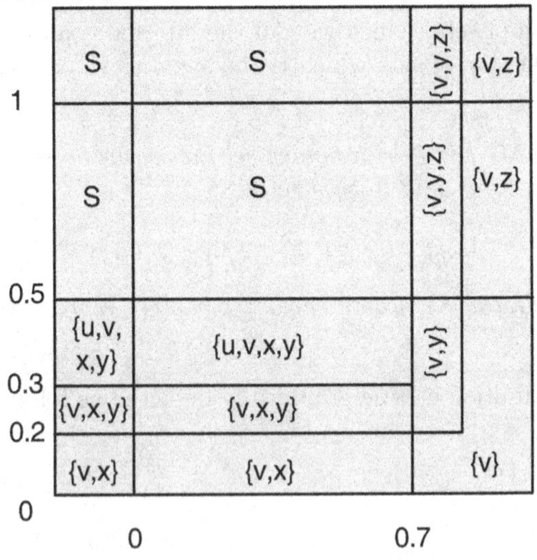

Recall that
$F_{0.4} = \{v, x\}$,
$F_{0.7} = \{v, x, y\}$,
$F_1 = \{u, v, x, y\}$,
$F_{1+\varepsilon} = S$.

Similarly,
$F^0 = S$,
$F^{0.3} = \{u, v, y, z\}$,
$F^{0.8} = \{v, y, z\}$,
$F^1 = \{v, z\}$.
Also note that
$G_{1+\varepsilon} = S$,
$G_{0.5} = \{u, x, y\}$,
$G_{0.3} = \{x, y\}$,
$G_{0.2} = \varnothing$.
Similarly,
$G^{0.3} = S$,
$G^{0.6} = \{u, v, x, z\}$,
$G^{0.7} = \{u, x\}$,
$G^1 = \varnothing$.
Now
$(F \cup G)_{0.2} = \{v, x\} = \{v, x\} \cup \varnothing = F_{0.2} \cup G_{0.2}$,
$(F \cup G)_{0.3} = \{v, x, y\} = \{v, x\} \cup \{x, y\} = F_{0.3} \cup G_{0.3}$,
$(F \cup G)_{0.4} = \{u, v, x, y\} = \{v, x, y\} \cup \{u, x, y\} = F_{0.4} \cup G_{0.4}$,
$(F \cup G)_{0.5} = \{u, v, x, y\} = \{v, x, y\} \cup \{u, x, y\} = F_{0.5} \cup G_{0.5}$,
$(F \cup G)_{0.7} = S = \{v, x, y\} \cup S = S$.
Further,
$(F \cup G)^1 = \{v, z\} = \{v, z\} \cup \varnothing = F^1 \cup G^1$.
$(F \cup G)^{0.8} = \{v, y, z\} = \{v, y, z\} \cup \varnothing = F^{0.8} \cup G^{0.8}$.
$(F \cup G)^{0.7} = S = \{v, y, z\} \cup \{u, x\} = F^{0.7} \cup G^{0.7}$.
It is worth noticing that the representation of F is
$F^\circ = (\{F_{0.4} = \{v, x\}, F_{0.7} = \{v, x, y\}, F_1 = \{u, v, x, y\}, F_{1+\varepsilon} = S\}$,
$\{F^0 = S, F^{0.3} = \{u, v, y, z\}, F^{0.8} = \{v, y, z\}, F^1 = \{v, z\}\})$.
Similarly,
$G^\circ = (\{G_{0.2} = \varnothing, G_{0.3} = \{x, y\}, G_{0.5} = \{u, x, y\}, G_{1+\varepsilon} = S\}$,
$\{G^{0.3} = S, G^{0.6} = \{u, v, x, z\}, G^{0.7} = \{u, x\}, G^1 = \varnothing\})$.
Now
$(F \cup G)^\circ = (\{(F \cup G)_{0.2} = \{v, x\}, (F \cup G)_{0.3} = \{v, x, y\}$,
$(F \cup G)_{0.5} = \{u, v, x, y\}, (F \cup G)_{0.7} = S\}$,
$\{(F \cup G)^{0.7} = S, (F \cup G)^{0.8} = \{v, y, z\}, (F \cup G)^1 = \{v, z\}\})$.
From the above computation, it is clear that $(F \cup G)^\circ$ *can be computed from* F° *and* G° *using the (regular) union operation. Thus the union operation seems to be precise.*

Theorem 5.2.2 *The binary operation union is precise.*

Proof. We shall prove that $(F \cup G)_s = F_s \cup G_s, (F \cup G)^t = F^t \cup G^t$ for $0 < s \leq 1 + \varepsilon, 0 \leq t \leq 1$. Let $x \in S$. Now $x \in (F \cup G)_s$ if and only if $l((F \cup G)(x)) < s$ if and only if $l(F(x) \cup G(x)) < s$ if and only if $l(F(x)) \wedge l(G(x)) < s$ if and only if $x \in F_s$ or $x \in G_s$ if and only if $x \in$

$F_s \cup G_s$. Similarly, $x \in (F \cup G)^t$ if and only if $u((F \cup G)(x)) \geq t$ if and only if $u(F(x) \cup G(x)) \geq t$ if and only if $u(F(x)) \vee u(G(x)) \geq t$ if and only if $x \in F^t$ or $x \in G^t$ if and only if $x \in F^t \cup G^t$. ■

Note that above theorem do not prove that $(F \cup G)_s^t = F_s^t \cup G_s^t$. In fact the result do not hold. Instead, we have the following:

Lemma 5.2.3 *Let F and G be two cisets. Then $(F \cup G)_s^t \supset F_s^t \cup G_s^t$ for $0 < s \leq 1 + \varepsilon, 0 \leq t \leq 1$.*

Proof. $(F \cup G)_s^t = (F \cup G)_s \cap (F \cup G)^t$
$= (F_s \cup G_s) \cap (F^t \cup G^t)$
$= (F_s \cap F^t) \cup (F_s \cap G^t) \cup (G_s \cap F^t) \cup (G_s \cap G^t)$
$= F_s^t \cup (F_s \cap G^t) \cup (G_s \cap F^t) \cup G_s^t$. ■

Example 5.2.4 *Consider the ciset F and G presented in Example 5.1.1 and in Example 5.1.5 respectively. Note that $F(y) = \langle 0.4, 0.8\rangle$ and $G(y) = \langle 0.2, 0.3\rangle$. Thus $(F \cup G)(y) = \langle 0.2, 0.8\rangle$. Therefore, we have the following:*
$y \in F^{0.7}, y \notin F_{0.3}, y \notin G^{0.7}, y \in G_{0.3}$.
Further, $y \in (F \cup G)_{0.3}^{0.7}$, $y \in (F \cup G)_{0.3}$ and $y \in (F \cup G)^{0.7}$.
Since, $y \notin F_{0.3}^{0.7}$ and $y \notin G_{0.3}^{0.7}$, $y \notin F_{0.3}^{0.7} \cup GG_{0.3}^{0.7}$. Thus $y \in (F \cup G)_{0.3}^{0.7} - (F_{0.3}^{0.7} \cup GG_{0.3}^{0.7})$.

Intersection

In this subsection, we show that the intersection operation is precise. We begin our discussion with the following motivating example.

Example 5.2.5 *Let F and G be cisets as in Example 5.1.1 and in Example 5.1.5 respectively.*
$(F \cap G)(u) = \langle 0.7, 0.3\rangle$,
$(F \cap G)(v) = \langle 0.5, 0.6\rangle$,
$(F \cap G)(x) = \langle 0.2, 0\rangle$,
$(F \cap G)(y) = \langle 0.4, 0.3\rangle$
$(F \cap G)(z) = \langle 1, 0.6\rangle$

FIGURE 5.4 Graph of $F \cap G$.

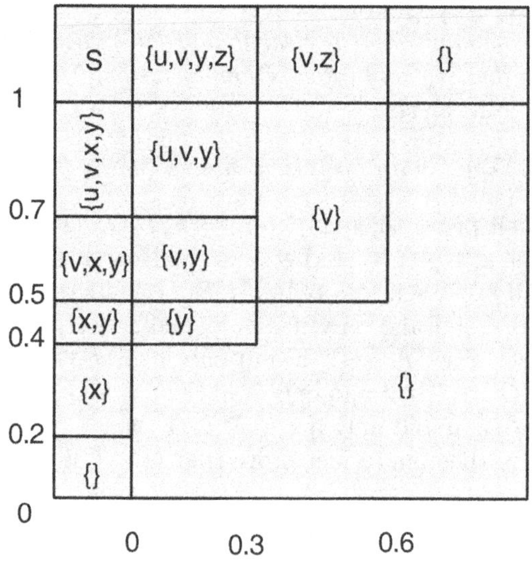

Now

$(F \cap G)_{0.2} = \varnothing = \{v, x\} \cap \varnothing = F_{0.2} \cap G_{0.2}$,
$(F \cap G)_{0.3} = \{x\} = \{v, x\} \cap \{x, y\} = F_{0.3} \cap G_{0.3}$,
$(F \cap G)_{0.4} = \{x\} = \{v, x\} \cap \{u, x, y\} = F_{0.4} \cap G_{0.4}$,
$(F \cap G)_{0.5} = \{x, y\} = \{v, x, y\} \cap \{u, x, y\} = F_{0.5} \cap G_{0.5}$,
$(F \cap G)_{0.7} = \{v, x, y\} = \{v, x, y\} \cap S = F_{0.7} \cap G_{0.7}$,
$(F \cap G)_1 = \{u, v, x, y\} = \{u, v, x, y\} \cap S = F_1 \cap G_1$,
$(F \cap G)_{1+\varepsilon} = S = S \cap S = F_{1+\varepsilon} \cap G_{1+\varepsilon}$,
Further,
$(F \cap G)^1 = \varnothing = \{v, z\} \cap \varnothing = F^1 \cap G^1$,
$(F \cap G)^{0.8} = \varnothing = \{v, y, z\} \cap \varnothing = F^{0.8} \cap G^{0.8}$,
$(F \cap G)^{0.7} = \varnothing = \{v, y, z\} \cap \{u, x\} = F^{0.7} \cap G^{0.7}$,
$(F \cap G)^{0.6} = \{v, z\} = \{v, y, z\} \cap \{u, v, x, z\} = F^{0.6} \cap G^{0.6}$,
$(F \cap G)^{0.3} = \{u, v, y, z\} = \{u, v, y, z\} \cap S = F^{0.3} \cap G^{0.3}$,
$(F \cap G)^0 = S = S \cap S = F^0 \cap G^0$.
It is worth noticing that the representation of F is
$F^\circ = (\{F_{0.4} = \{v, x\}, F_{0.7} = \{v, x, y\}, F_1 = \{u, v, x, y\}, F_{1+\varepsilon} = S\}$,
$\{F^0 = S, F^{0.3} = \{u, v, y, z\}, F^{0.8} = \{v, y, z\}, F^1 = \{v, z\}\})$.
Similarly,
$G^\circ = (\{G_{0.2} = \varnothing, G_{0.3} = \{x, y\}, G_{0.5} = \{u, x, y\}, G_{1+\varepsilon} = S\}$,
$\{G^{0.3} = S, G^{0.6} = \{u, v, x, z\}, G^{0.7} = \{u, x\}, G^1 = \varnothing\})$.
Now
$(F \cap G)^\circ = (\{(F \cap G)_{0.2} = \varnothing, (F \cap G)_{0.4} = \{x\}, (F \cap G)_{0.5} = \{x, y\}$,
$(F \cap G)_{0.7} = \{v, x, y\}, (F \cap G)_1 = \{u, v, x, y\}$,

$(F \cap G)_{1+\varepsilon} = S\}, \{(F \cap G)^{0.6} = \{v, z\},$
$(F \cap G)^{0.3} = \{u, v, y, z\}, (F \cap G)^{0} = S\})$.
From the above computation, it is clear that $(F \cap G)^\circ$ can be computed from F° and G° using the (regular) intersection operation. Thus the intersection operation seems to be precise.

Theorem 5.2.6 *The binary operation intersection is precise.*

Proof. We shall prove that $(F \cap G)_s = F_s \cap G_s, (F \cap G)^t = F^t \cap G^t$ for $0 < s \leq 1 + \varepsilon, 0 \leq t \leq 1$. Let $x \in S$. Now $x \in (F \cap G)_s$ if and only if $l((F \cap G)(x)) < s$ if and only if $l(F(x) \cap G(x)) < s$ if and only if $l(F(x)) \vee l(G(x)) < s$ if and only if $x \in F_s$ and $x \in G_s$ if and only if $x \in F_s \cap G_s$. Similarly, $x \in (F \cap G)^t$ if and only if $u((F \cap G)(x)) \geq t$ if and only if $u(F(x) \cap G(x)) \geq t$ if and only if $u(F(x)) \wedge u(G(x)) \geq t$ if and only if $x \in F^t$ and $x \in G^t$ if and only if $x \in F^t \cap G^t$. ■

Note that above theorem in fact prove that $(F \cap G)_s^t = F_s^t \cap G_s^t$.

Corollary 5.2.7 *Let F and G be two cisets. Then $(F \cap G)_s^t = F_s^t \cap G_s^t$ for* $0 < s \leq 1 + \varepsilon, 0 \leq t \leq 1$.

Proof. $(F \cap G)_s^t = (F \cap G)_s \cap (F \cap G)^t$
$= (F_s \cap G_s) \cap (F^t \cap G^t)$
$= (F_s \cap F^t) \cap (G_s \cap G^t)$
$= F_s^t \cap G_s^t$. ■

Negation

In this subsection, we show that the negation operation is precise. We begin our discussion with the following motivating example.

Example 5.2.8 *Let F be a ciset as given in Example 5.1.1.*
$F(u) = \langle 0.7, 0.3 \rangle,$
$F(v) = \langle 0, 1 \rangle,$
$F(x) = \langle 0, 0 \rangle,$
$F(y) = \langle 0.4, 0.8 \rangle,$
$F(z) = \langle 1, 1 \rangle.$
$-F(u) = \langle 0.3, 0.7 \rangle,$
$-F(v) = \langle 1, 0 \rangle,$
$-F(x) = \langle 0, 0 \rangle,$
$-F(y) = \langle 0.8, 0.4 \rangle,$
$-F(z) = \langle 1, 1 \rangle.$

Now
$(-F)_0 = \varnothing = S - S = S - F^0 = (F^0)^c,$
$(-F)_{0.3} = \{x\} = S - \{u, v, y, z\} = (F^{0.3})^c,$
$(-F)_{0.8} = \{u, x\} = S - \{v, y, z\} = (F^{0.8})^c,$

FIGURE 5.5 Graph of $-F$.

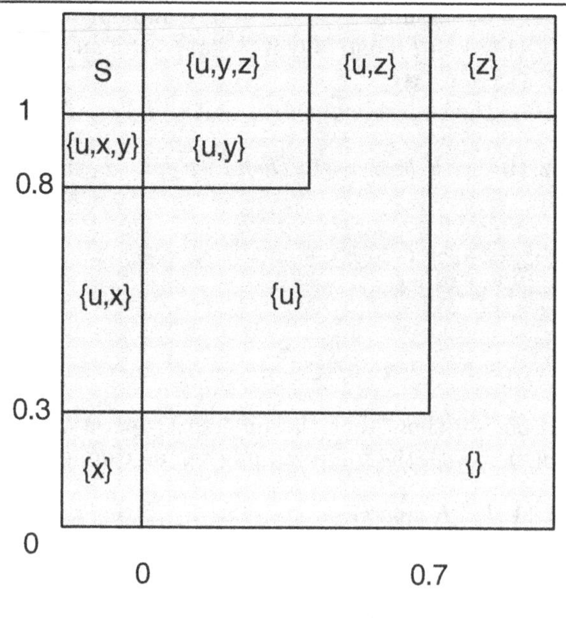

$(-F)_1 = \{u, x, y\} = S - \{v, z\} = (F^1)^c$,
$(-F)_{1+\varepsilon} = S = S - \varnothing = (F^{1+\varepsilon})^c$.
Further,
$(-F)^0 = S = S - \varnothing = S - F_0 = (F_0)^c$,
$(-F)^{0.4} = \{u, y, z\} = S - \{v, x\} = (F_{0.4})^c$,
$(-F)^{0.7} = \{u, z\} = S - \{v, x, y\} = (F_{0.7})^c$,
$(-F)^1 = \{z\} = S - \{u, v, x, y\} = (F_1)^c$,
$(-F)^{1+\varepsilon} = \varnothing = S - S = (F_{1+\varepsilon})^c$.
It is worth noticing that the representation of F is
$F^\circ = (\{F_{0.4} = \{v, x\}, F_{0.7} = \{v, x, y\}, F_1 = \{u, v, x, y\}, F_{1+\varepsilon} = S\}$,
$\{F^0 = S, F^{0.3} = \{u, v, y, z\}, F^{0.8} = \{v, y, z\}, F^1 = \{v, z\}\})$.
Now
$(-F)^\circ = (\{(-F)^0 = S, (-F)^{0.4} = \{u, y, z\}, (-F)^{0.7} = \{u, z\}, (-F)^1 = \{z\}$,
$(-F)^{1+\varepsilon} = \varnothing\}, \{(-F)_0 = \varnothing, (-F)_{0.3} = \{x\}, (-F)_{0.8} = \{u, x\}$,
$(-F)_1 = \{u, x, y\}, (-F)_{1+\varepsilon} = S\})$.
From the above computation, it is clear that $(-F)^\circ$ can be computed from F° using the (regular) complement operation. Thus the negation operation seems to be precise.

Theorem 5.2.9 *The binary operation negation is precise.*

Proof. We shall prove that $(-F)_s = S - F^s, (-F)^t = S - F_t$ for $0 < s \leq 1 + \varepsilon, 0 \leq t \leq 1$. Let $x \in S$. Now $x \in (-F)_s$ if and only if $l((-F)(x)) < s$ if and only if $l(-F(x)) < s$ if and only if $u(F(x)) < s$ if and only if $x \in (F^s)^c$ if and only if $x \in S - F^s$. Similarly, $x \in (-F)^t$ if and only if $u((-F)(x)) \geq t$ if and only if $u(-F(x)) \geq t$ if and only if $l(F(x)) \geq t$ if and only if $x \in (F_t)^c$ if and only if $x \in S - F_t$. ■

Note that above theorem do not prove that $(-F)^t_s = S - F^t_s$.

Corollary 5.2.10 *Let F be a ciset. Then $(-F)^t_s = (F_t \cup F^s)^c$ for $0 < s \leq 1 + \varepsilon, 0 \leq t \leq 1$.*

Proof. $(-F)^t_s = (-F)^t \cap (-F)_s$
$(F_t)^c \cap (F^s)^c = (F_t \cup F^s)^c$ ■

Difference

In this subsection, we show that the difference operation is precise. We begin our discussion with the following motivating example.

Example 5.2.11 *Let F and G be cisets as in Example 5.1.1 and in Example 5.1.5 respectively.*
$(F - G)(u) = \langle 0.7, 0.3 \rangle$,
$(F - G)(v) = \langle 0.6, 0.5 \rangle$,
$(F - G)(x) = \langle 0.7, 0 \rangle$,
$(F - G)(y) = \langle 0.4, 0.2 \rangle$
$(F - G)(z) = \langle 1, 0.5 \rangle$

FIGURE 5.6 Graph of $F - G$.

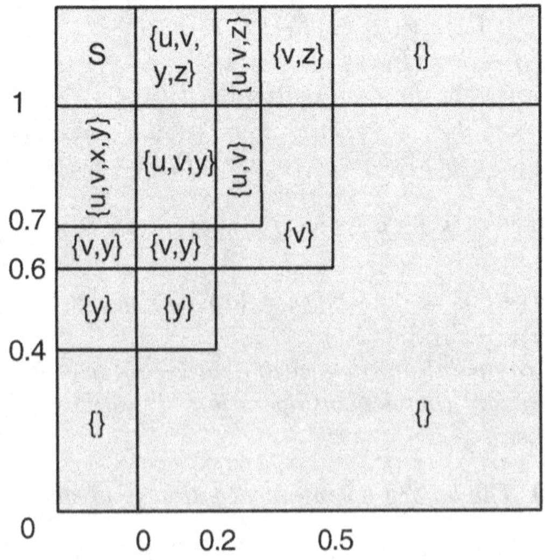

Now
$(F-G)_{0.2} = \varnothing = \{v,x\} - \{u,v,x,z\} = F_{0.2} - G^{0.2}$,
$(F-G)_{0.3} = \varnothing = \{v,x\} - \{u,v,x,z\} = F_{0.3} - G^{0.2}$,
$(F-G)_{0.4} = \varnothing = \{v,x\} - \{u,v,x,z\} = F_{0.4} - G^{0.4}$,
$(F-G)_{0.6} = \{y\} = \{v,x,y\} - \{u,v,x,z\} = F_{0.6} - G^{0.6}$,
$(F-G)_{0.7} = \{v,y\} = \{v,x,y\} - \{u,x\} = F_{0.7} - G^{0.7}$,
$(F-G)_{1} = \{u,v,x,y\} = \{u,v,x,y\} - \varnothing = F_{1} - G^{1}$,
$(F-G)_{1+\varepsilon} = S = S - \varnothing = F_{1+\varepsilon} - G^{1+\varepsilon}$.
Further,
$(F-G)^{1} = \varnothing = \{v,z\} - S = F^{1} - G_{1}$,
$(F-G)^{0.8} = \varnothing = \{v,y,z\} - S = F^{0.8} - G_{0.8}$,
$(F-G)^{0.7} = \varnothing = \{v,y,z\} - S = F^{0.7} - G_{0.7}$,
$(F-G)^{0.5} = \{v,z\} = \{v,y,z\} - \{u,x,y\} = F^{0.5} - G_{0.5}$,
$(F-G)^{0.3} = \{u,v,z\} = \{u,v,y,z\} - \{x,y\} = F^{0.3} - G_{0.3}$,
$(F-G)^{0.2} = \{u,v,y,z\} = \{u,v,y,z\} - \varnothing = F^{0.2} - G_{0.2}$.
$(F-G)^{0} = S = S - \varnothing = F^{0} - G_{0}$.
It is worth noticing that the representation of F is
$F^\circ = (\{F_{0.4} = \{v,x\}, F_{0.7} = \{v,x,y\}, F_1 = \{u,v,x,y\}, F_{1+\varepsilon} = S\}$,
$\{F^0 = S, F^{0.3} = \{u,v,y,z\}, F^{0.8} = \{v,y,z\}, F^1 = \{v,z\}\})$.
Similarly,
$G^\circ = (\{G_{0.2} = \varnothing, G_{0.3} = \{x,y\}, G_{0.5} = \{u,x,y\}, G_{1+\varepsilon} = S\}$,
$\{G^{0.3} = S, G^{0.6} = \{u,v,x,z\}, G^{0.7} = \{u,x\}, G^1 = \varnothing\})$.
Now
$(F-G)^\circ = (\{(F-G)_{0.4} = \varnothing, (F-G)_{0.6} = \{y\}, (F-G)_{0.7} = \{v,y\}$,
$(F-G)_1 = \{u,v,x,y\}, (F-G)_{1+\varepsilon} = S\}, \{(F-G)^0 = S$,
$(F-G)^{0.2} = \{u,v,y,z\}, (F-G)^{0.3} = \{u,v,z\}$,
$(F-G)^{0.5} = \{v,z\}\})$.
From the above computation, it is clear that $(F-G)^\circ$ can be computed from F° and G° using the (regular) difference operation. Thus the difference operation seems to be precise.

Theorem 5.2.12 *The binary operation difference is precise.*

Proof. We shall prove that $(F-G)_s = F_s - G^s, (F-G)^t = F^t - G_t$ for $0 < s \le 1+\varepsilon, 0 \le t \le 1$. Let $x \in S$. Now $x \in (F-G)_s$ if and only if $l((F-G)(x)) < s$ if and only if $l(F(x)) \cap u(G(x)) < s$ if and only if $l(F(x)) \vee u(G(x)) < s$ if and only if $x \in F_s$ and $x \in G^s$ if and only if $x \in F_s - G^s$. Similarly, $x \in (F-G)^t$ if and only if $u((F-G)(x)) \ge t$ if and only if $u(F(x) - G(x)) \ge t$ if and only if $u(F(x)) \wedge l(G(x)) \ge t$ if and only if $x \in F^t$ and $x \in G_t$ if and only if $x \in F^t - G_t$. ■

Note that above theorem do not prove that $(F-G)_s^t = F_s^t - G_s^t$. However, we have the following result.

Corollary 5.2.13 *Let F and G be two cisets. Then $(F-G)_s^t = F_s^t \cap (-G_s^t)$ for $0 < s \le 1+\varepsilon, 0 \le t \le 1$.*

Proof. $(F - G)_s^t = (F - G)_s \cap (F - G)^t$
$= (F_s - G^s) \cap (F^t - G_t)$
$= (F_s \cap (G^s)^c) \cap (F^t \cap (G_t)^c)$
$= F_s \cap (G^s)^c \cap F^t \cap (G_t)^c$
$= F_s^t \cap (G^s \cup G_t)^c$
$= F_s^t \cap (-G_s^t)$. ■

Example 5.2.14 *Let F and G be cisets as in Example 5.1.1 and in Example 5.1.5 respectively.*
$(F - G)_{.65}^{.45} = \{v\}$.
$(F - G)_{.65} \cap (F - G)^{.45} = \{v, y\} \cap \{v, z\} = \{v\}$.
$F_{.65} - G^{.65} = F_{.65} \cap (G^{.65})^c = \{v, x, y\} \cap \{v, y, z\}$
$F^{.45} - G_{.45} = F^{.45} \cap (G_{.45})^c = \{v, y, z\} \cap \{v, z\}$.

Cartesian product

In this subsection, we show that the Cartesian product operation is precise. We begin our discussion with the following motivating example.

Example 5.2.15 *Let F and G be cisets as in Example 5.1.1 and in Example 5.1.5 respectively.*
$(F \times G)(u, u) = \langle 0.7, 0.3\rangle, (F \times G)(u, v) = \langle 0.7, 0.3\rangle,$
$(F \times G)(u, x) = \langle 0.7, 0.3\rangle, (F \times G)(u, y) = \langle 0.7, 0.3\rangle,$
$(F \times G)(u, z) = \langle 0.7, 0.3\rangle, (F \times G)(v, u) = \langle 0.3, 0.7\rangle,$
$(F \times G)(v, v) = \langle 0.5, 0.6\rangle, (F \times G)(v, x) = \langle 0.2, 0.7\rangle,$
$(F \times G)(v, y) = \langle 0.2, 0.3\rangle, (F \times G)(v, z) = \langle 0.5, 0.6\rangle,$
$(F \times G)(x, u) = \langle 0.3, 0\rangle, (F \times G)(x, v) = \langle 0.5, 0\rangle,$
$(F \times G)(x, x) = \langle 0.2, 0\rangle, (F \times G)(x, y) = \langle 0.2, 0\rangle,$
$(F \times G)(x, z) = \langle 0.5, 0\rangle, (F \times G)(y, u) = \langle 0.4, 0.7\rangle,$
$(F \times G)(y, v) = \langle 0.5, 0.6\rangle, (F \times G)(y, x) = \langle 0.4, 0.7\rangle,$
$(F \times G)(y, y) = \langle 0.4, 0.3\rangle, (F \times G)(y, z) = \langle 0.5, 0.6\rangle,$
$(F \times G)(z, u) = \langle 1, 0.7\rangle, (F \times G)(z, v) = \langle 1, 0.6\rangle,$
$(F \times G)(z, x) = \langle 1, 0.7\rangle, (F \times G)(z, u) = \langle 1, 0.3\rangle,$
$(F \times G)(z, u) = \langle 1, 0.6\rangle.$

Now
$(F \times G)_{0.2} = \varnothing = \{v, x\} \times \varnothing = F_{0.2} \times G_{0.2},$
$(F \times G)_{0.3} = \{v, x\} \times \{x, y\} = F_{0.3} \times G_{0.3},$
$(F \times G)_{0.4} = \{v, x\} \times \{u, x, y\} = F_{0.4} \times G_{0.4},$
$(F \times G)_{0.5} = \{v, x, y\} \times \{u, x, y\} = F_{0.5} \times G_{0.5},$
$(F \times G)_{0.7} = \{v, x, y\} \times S = F_{0.7} \times G_{0.7},$
$(F \times G)_1 = \{u, v, x, y\} \times S = F_1 \times G_1,$
$(F \times G)_{1+\varepsilon} = S \times S = F_{1+\varepsilon} \times G_{1+\varepsilon},$
Further,
$(F \times G)^1 = \varnothing = \{v, z\} \times \varnothing = F^1 \times G^1,$
$(F \times G)^{0.7} = \{v, y, z\} \times \{u, x\} = F^{0.7} \times G^{0.7},$

$(F \times G)^{0.6} = \{v, y, z\} \times \{u, v, x, z\} = F^{0.6} \times G^{0.6}$,
$(F \times G)^{0.3} = \{u, v, y, z\} \times S = F^{0.3} \times G^{0.3}$,
$(F \times G)^{0} = S \times S = F^{0} \times G^{0}$.
It is worth noticing that the representation of F is
$F^{\circ} = (\{F_{0.4} = \{v, x\}, F_{0.7} = \{v, x, y\}, F_1 = \{u, v, x, y\}, F_{1+\varepsilon} = S\}$,
$\{F^0 = S, F^{0.3} = \{u, v, y, z\}, F^{0.8} = \{v, y, z\}, F^1 = \{v, z\}\})$.
Similarly,
$G^{\circ} = (\{G_{0.2} = \varnothing, G_{0.3} = \{x, y\}, G_{0.5} = \{u, x, y\}, G_{1+\varepsilon} = S\}$,
$\{G^{0.3} = S, G^{0.6} = \{u, v, x, z\}, G^{0.7} = \{u, x\}, G^1 = \varnothing\})$.
Now
$(F \times G)^{\circ} = (\{(F \times G)_{0.2} = \varnothing, (F \times G)_{0.3} = \{v, x\} \times \{x, y\}$,
$(F \times G)_{0.4} = \{v, x\} \times \{u, x, y\}$,
$(F \times G)_{0.5} = \{v, x, y\} \times \{u, x, y\}$,
$(F \times G)_{0.7} = \{v, x, y\} \times S, (F \times G)_1 = \{u, v, x, y\} \times S$,
$(F \times G)_{1+\varepsilon} = S \times S\}, \{(F \times G)^0 = S \times S$,
$(F \times G)^{0.3} = \{u, v, y, z\} \times S$,
$(F \times G)^{0.6} = \{v, y, z\} \times \{u, v, x, z\}$,
$(F \times G)^{0.7} = \{v, y, z\} \times \{u, x\}, (F \times G)^1 = \varnothing\})$.
From the above computation, it is clear that $(F \times G)^{\circ}$ can be computed from F° and G° using the (regular) Cartesian product operation. Thus the Cartesian product operation seems to be precise.

Theorem 5.2.16 *The binary operation Cartesian product is precise.*

Proof. We shall prove that $(F \times G)_s = F_s \times G_s, (F \times G)^t = F^t \times G^t$ for $0 < s \leq 1 + \varepsilon, 0 \leq t \leq 1$. Let $x, y \in S$. Now $(x, y) \in (F \times G)_s$ if and only if $l((F \times G)(x, y)) < s$ if and only if $l(F(x) \cap G(y)) < s$ if and only if $l(F(x)) \vee l(G(y)) < s$ if and only if $x \in F_s$ and $y \in G_s$ if and only if $(x, y) \in F_s \times G_s$. Similarly, $(x, y) \in (F \times G)^t$ if and only if $u((F \times G)(x, y)) \geq t$ if and only if $u(F(x) \cap G(y)) \geq t$ if and only if $u(F(x)) \wedge u(G(y)) \geq t$ if and only if $x \in F^t$ and $y \in G^t$ if and only if $(x, y) \in F^t \times G^t$. ■

Note that above theorem in fact prove that $(F \times G)_s^t = F_s^t \times G_s^t$.

Corollary 5.2.17 *Let F and G be two cisets. Then $(F \cap G)_s^t = F_s^t \cap G_s^t$ for $0 < s \leq 1 + \varepsilon, 0 \leq t \leq 1$.*

Proof. $(F \times G)_s^t = (F \times G)_s \cap (F \times G)^t$
$= (F_s \times G_s) \cap (F^t \times G^t)$
$= (F_s \cap F^t) \times (G_s \cap G^t)$
$= F_s^t \times G_s^t$. ■

Since a ciset relation is a ciset, above results carry over to ciset relations. Thus we have the following

Theorem 5.2.18 *The binary ciset relational operations selection, projection, union, intersection, Cartesian product, join, equijoin and theta-join are precise.*

Proof. Proof that ciset relational operations selection and projection are precise is clear due to the fact that ciset relational operations selection and projection are the same as the corresponding relational operations. Proof that ciset relational operations union, intersection, difference and Cartesian product are precise follows from the fact that a ciset relation is indeed a ciset and we have already proved that ciset operations union, intersection, difference and Cartesian product are precise. Proof that ciset relational operations join, equijoin, theta-join and divide are precise, follows from the observation that any one of these operations can be implemented using selection, projection, union, intersection, difference and Cartesian product and they are already shown to be precise. ■

In the next Chapter we present additional operations on cisets. These operations are not necessary to implement ciset database and hence we did not introduce them so far. However, we believe they may be useful in many other applications such as expert systems, knowledge-based systems, data mining and so on.

6

ADDITIONAL CISET OPERATIONS

In this chapter we introduce four additional operations on ciset and study various properties of those operations. We also introduce a new partial order $\leq$ on $\mathfrak{C}$.

6.1 Confidence Index Operations

Two confidence indexes $a_i = \langle \alpha_i, \beta_i \rangle, i = 1, 2$ are equal if and only if $l(a_1) = l(a_2)$ and $u(a_1) = u(a_2)$. We now proceed to define a new partial order on confidence indexes. Confidence index $a_1 < a_2$ if $l(a_1) \leq l(a_2)$ and $u(a_1) < u(a_2)$ or $l(a_1) < l(a_2)$ and $u(a_1) \leq u(a_2)$. Of course $a_1 \leq a_2$ if and only if either $a_1 < a_2$ or $a_1 = a_2$. Further, $a_1 > a_2$ if and only if $a_2 < a_1$ and $a_1 \geq a_2$ if and only if $a_2 \leq a_1$.

We now proceed to introduce four operations. They are three binary operations *c-union* ($\sqcup$), *c-intersection* ($\sqcap$),*c-difference* ($\neg$); and one unary operation *c-negation* ($\neg$). Let $a_1 = \langle \beta_1, \alpha_1 \rangle, a_2 = \langle \beta_2, \alpha_2 \rangle$ be any two confidence indexes. Then

$$a_1 \sqcup a_2 = \langle \alpha_1 \vee \alpha_2, \beta_1 \vee \beta_2 \rangle,$$

$$a_1 \sqcap a_2 = \langle \alpha_1 \wedge \alpha_2, \beta_1 \wedge \beta_2 \rangle,$$

$$a_1 \neg a_2 = \begin{cases} a_1 & \text{if } a_1 > a_2 \\ \langle 0, u(a_1)\rangle & \text{if } u(a_1) > u(a_2) \text{ and } l(a_1) \leq l(a_2) \\ \langle l(a_1), 0\rangle & \text{if } l(a_1) > l(a_2) \text{ and } u(a_1) \leq u(a_2) \\ \langle 0, 0\rangle & \text{if } a_1 \leq a_2 \end{cases},$$

and

$$\neg a_1 = \langle 1 - \alpha_1, 1 - \beta_1 \rangle.$$

Example 6.1.1 *Let* $a_1 = \langle 0.3, 0.7\rangle, a_2 = \langle 0.5, 0.7\rangle, a_3 = \langle 0.4, 0.6\rangle, a_4 = \langle 0.2, 0.8\rangle$. *Then we have the following:*
$a_1 < a_2 > a_3$;
$a_1 \sqcup a_2 = a_2$;
$a_2 \sqcap a_3 = a_3$;
$a_5 = a_3 \sqcup a_4 = \langle 0.4, 0.8\rangle > a_3$ *and* $a_5 > a_4$;
$a_6 = a_3 \sqcap a_4 = \langle 0.2, 0.6\rangle < a_3$ *and* $a_6 < a_4$;
$\neg a_2 = \langle 0.5, 0.3\rangle$;
$\langle 0, 1\rangle \neg a_2 = \langle 0, 1\rangle$;
$\neg(a_3 \sqcup a_4) = \neg\langle 0.4, 0.8\rangle = \langle 0.6, 0.2\rangle = \langle 0.6, 0.4\rangle \sqcap \langle 0.8, 0.2\rangle = (\neg a_3) \sqcap (\neg a_4)$;
and
$\neg(a_3 \sqcap a_4) = \neg\langle 0.2, 0.6\rangle = \langle 0.8, 0.4\rangle = \langle 0.6, 0.4\rangle \sqcup \langle 0.8, 0.2\rangle = (\neg a_3) \sqcup (\neg a_4)$.

Example 6.1.2 *Consider two statements: (A) The Facility X in country ABC is used to produce biological weapons and (B) The Facility X in country ABC is* **not** *used to produce biological weapons. Clearly, (A) and (B) are opposite statements. Since the confidence index of (A) is* $\langle 0.3, 0.6\rangle$, *it is reasonable to assign* $\langle 0.7, 0.4\rangle$ *as the confidence index of (B).*

Definition 6.1.3 *Let A and B be two statements such that A is the c-negation of B in the logical sense. If a is the confidence index of statement A, then the confidence index of the statement B is $\neg a$.*

The following result is quite clear and hence we state it without proof.

Proposition 6.1.4 *Let a be a confidence index. Then a is equal to $\neg a$ if and only if* $l(a) = u(a) = \langle 0.5, 0.5\rangle$.

Theorem 6.1.5 *The set of all confidence indexes $\mathfrak{C}$ is closed under c-union, c-intersection, c-difference and c-negation operations.*

Theorem 6.1.6 *Let $a_i \in \mathfrak{C}$ for $i = 1, 2, 3$. Then we have the following:*

1. $a_1 \sqcup a_2 = a_2 \sqcup a_1$,

2. $a_1 \sqcap a_2 = a_2 \sqcap a_1$,

3. $a_1 \sqcup \langle 0, 0\rangle = a_1$,

4. $a_1 \sqcap \langle 1,1\rangle = a_1$,

5. $a_1 \sqcup \langle 1,1\rangle = \langle 1,1\rangle$,

6. $a_1 \sqcap \langle 0,0\rangle = \langle 0,0\rangle$,

7. $a_1 \sqcup a_1 = a_1$,

8. $a_1 \sqcap a_1 = a_1$,

9. $a_1 \sqcup (a_2 \sqcup a_3) = (a_1 \sqcup a_2) \sqcup a_3$,

10. $a_1 \sqcap (a_2 \sqcap a_3) = (a_1 \sqcap a_2) \sqcap a_3$,

11. $a_1 \sqcup (a_2 \sqcap a_3) = (a_1 \sqcup a_2) \sqcap (a_1 \sqcup a_3)$,

12. $a_1 \sqcap (a_2 \sqcup a_3) = (a_1 \sqcap a_2) \sqcup (a_1 \sqcap a_3)$,

13. $\neg(a_1 \sqcup a_2) = (\neg a_1) \sqcap (\neg a_2)$,

14. $\neg(a_1 \sqcap a_2) = (\neg a_1) \sqcup (\neg a_2)$,

15. $(\neg(\neg a_1)) = a_1$,

16. *if* $a_1 \leq a_2$ *then* $a_1 \sqcup a_2 = a_2$ *and* $a_1 \sqcap a_2 = a_1$,

17. *If* $a_1 \leq a_3, a_2 \leq a_4$ *then* $a_1 \sqcup a_2 \leq a_3 \sqcup a_4$,

18. *If* $a_1 \leq a_3, a_2 \leq a_4$ *then* $a_1 \sqcap a_2 \leq a_3 \sqcap a_4$.

■

It is worthwhile to note that the properties $a_1 \sqcap (\neg a_1) = \langle 0,0\rangle$ and $a_1 \sqcup (\neg a_1) = \langle 1,1\rangle$ do not hold in general. In logic, the former property is known as the *law of contradiction* and the latter property is called the *law of excluded middle.*

The reader may note that (13) and (14) above prove that *DeMorgan's Laws* do hold.

Example 6.1.7 *Let* $a_1 = \langle 0.5, 0.7\rangle$ *and* $a_2 = \langle 0.6, 0.3\rangle$. *Then we have the following:*
$\neg a_1 = \langle 0.5, 0.3\rangle, \neg a_2 = \langle 0.4, 0.7\rangle$,
$a_1 \neg a_1 = \langle 0,0\rangle$,
$a_2 \neg a_2 = \langle 0,0\rangle$,
$a_1 \neg a_2 = \langle 0.5, 0.7\rangle \neg \langle 0.6, 0.3\rangle = \langle 0, 0.7\rangle$,
$a_2 \neg a_1 = \langle 0.6, 0.3\rangle \neg \langle 0.5, 0.7\rangle = \langle 0.6, 0\rangle$,
$a_1 \sqcup a_2 = \langle 0.5, 0.7\rangle \sqcup \langle 0.6, 0.3\rangle = \langle 0.6, 0.7\rangle$,
$a_1 \sqcap a_2 = \langle 0.5, 0.7\rangle \sqcap \langle 0.6, 0.3\rangle = \langle 0.5, 0.3\rangle$.
$\langle 0,0\rangle = (a_1 \neg a_2) \sqcap (a_2 \neg a_1) \leq a_1 \sqcap a_2 \leq (a_1 \neg a_2) \sqcup (a_2 \neg a_1) = a_1 \sqcup a_2$.

Proposition 6.1.8 *Let* $a_i \in \mathfrak{C}$ *for* $i = 1,2,3$. *We have the following:*

1. $\langle 0,1\rangle\neg a_1 = \langle 0,1\rangle$ if $u(a_1) < 1$.
2. $\langle 1,0\rangle\neg a_1 = \langle 1,0\rangle$ if $l(a_1) < 1$.
3. $a_1\neg\langle 1,0\rangle = \langle 0, u(a_1)\rangle$,
4. $a_1\neg\langle 0,1\rangle = \langle l(a_1), 0\rangle$,
5. $a_1\neg a_1 = \langle 0,0\rangle$,
6. $(a_1 \sqcup a_2)\neg a_3 = (a_1\neg a_3) \sqcup (a_2\neg a_3)$,
7. $a_1\neg(a_2 \sqcap a_3) = (a_1\neg a_2) \sqcup (a_1\neg a_3)$,
8. $(a_1 \sqcap a_2)\neg a_3 = (a_1\neg a_3) \sqcap (a_2\neg a_3)$,
9. $a_1\neg(a_2 \sqcup a_3) = (a_1\neg a_2) \sqcap (a_1\neg a_3)$.

Proof. Proof of first five statements are easy to follow and hence omitted.

6. Without loss of generality, assume that $u(a_1). \geq u(a_2)$.First observe that the only possible values of $u((a_1 \sqcup a_2)\neg a_3)$ are 0 and $u(a_1)$.Similarly, the only possible values of $u((a_1\neg a_3) \sqcup (a_2\neg a_3))$ are 0 and $u(a_1)$. Now $u((a_1 \sqcup a_2)\neg a_3) = 0$ if and only if $u(a_3) \geq u(a_1)$. That is, $u((a_1 \sqcup a_2)\neg a_3) = 0$ if and only if $u(a_1\neg a_3) = 0$. Thus $u((a_1 \sqcup a_2)\neg a_3) = 0$ if and only if $u((a_1\neg a_3) \sqcup (a_2\neg a_3)) = 0$. Similar is the case for lower indexes and hence the result.

7. Without loss of generality, assume that $u(a_2). \geq u(a_3)$.First observe that the only possible values of $u(a_1\neg(a_2 \sqcap a_3))$ are 0 and $u(a_1)$.Similarly, the only possible values of $u((a_1\neg a_2) \sqcup (a_1\neg a_3))$ are 0 and $u(a_1)$. Now $u(a_1\neg(a_2 \sqcap a_3)) = 0$ if and only if $u(a_3) \geq u(a_1)$. That is, $u(a_1\neg(a_2 \sqcap a_3)) = 0$ if and only if $u(a_1\neg a_3) = 0$ and $u(a_1\neg a_2) = 0$ Thus $u(a_1\neg(a_2 \sqcap a_3)) = 0$ if and only if $u((a_1\neg a_3) \sqcup (a_2\neg a_3)) = 0$.Similar is the case for lower indexes and hence the result.

8. Without loss of generality, assume that $u(a_1). \leq u(a_2)$.First observe that the only possible values of $u((a_1 \sqcap a_2)\neg a_3)$ are 0 and $u(a_1)$.Similarly, the only possible values of $u((a_1\neg a_3) \sqcap (a_2\neg a_3))$ are 0 and $u(a_1)$. Now $u((a_1 \sqcap a_2)\neg a_3) = 0$ if and only if $u(a_3) \geq u(a_1)$. That is, $u((a_1 \sqcap a_2)\neg a_3) = 0$ if and only if $u(a_1\neg a_3) = 0$. Thus $u((a_1 \sqcap a_2)\neg a_3) = 0$ if and only if $u((a_1\neg a_3) \sqcap (a_2\neg a_3)) = 0$. Similar is the case for lower indexes and hence the result.

9. Without loss of generality, assume that $u(a_2). \leq u(a_3)$.First observe that the only possible values of $u(a_1\neg(a_2 \sqcup a_3))$ are 0 and $u(a_1)$.Similarly, the only possible values of $u((a_1\neg a_2) \sqcap (a_1\neg a_3))$ are 0 and $u(a_1)$. Now $u(a_1\neg(a_2 \sqcup a_3)) = 0$ if and only if $u(a_3) \geq u(a_1)$. That is, $u(a_1\neg(a_2 \sqcap a_3)) = 0$ if and only if $u(a_1\neg a_3) = 0$. Thus $u(a_1\neg(a_2 \sqcup a_3)) = 0$ if and only if $u((a_1\neg a_3) \sqcap (a_2\neg a_3)) = 0$.Similar is the case for lower indexes and hence the result.

■.

We say two cisets F and G on a set S are equal, and write $F = G$, if $F(x) = G(x)$ for all $x \in S$.

Definition 6.1.9 *Let F and G be two cisets on a set S such that $F(x) \leq G(x)$ for all $x \in S$, then F is said to be* c-subset *of G and G is said to be a* c-superset *of F. If F is a subset of G and there exists at least one $x \in S$ such that $F(x) < G(x)$ then F is said to be* proper c-subset *of G and G is said to be a* proper c-superset *of F.*

If F is a c-subset of G, F is said to be c-contained in G. Symbolically, we write $F \sqsubseteq G$. If F is a proper c-subset of G, then we say F is strictly c-contained in G, denoted by $F \sqsubset G$.

Proposition 6.1.10 *Let F, G, H be cisets on a set S. Then*

1. $F \sqsubseteq F$,
2. If $F \sqsubseteq G$ and $G \sqsubseteq H$, then $F \sqsubseteq H$,
3. If $F \sqsubseteq G$ and $G \sqsubset H$, then $F \sqsubset H$,
4. If $F \sqsubseteq G$ and $F \not\sqsubseteq H$, then $G \not\sqsubseteq H$, where $\not\sqsubseteq$ means "is not c-contained in,"
5. $F = G$ if and only if $F \sqsubseteq G$ and $G \sqsubseteq F$.

Proof.

1. For all $x \in S, F(x) \leq F(x)$. Thus $F \sqsubseteq F$.
2. For all $x \in S, F(x) \leq G(x) \leq H(x)$. Thus $F \sqsubseteq H$.
3. For all $x \in S, F(x) \leq G(x) < H(x)$. Thus $F \sqsubset H$.
4. For all $x \in S, F(x) \leq G(x)$ and there exists an element $y \in S$ such that $F(y) > H(y)$. Thus $H(y) < F(y) \leq G(x)$. Thus $G \not\sqsubseteq H$.
5. $F = G \Leftrightarrow$ for all $x \in S, F(x) = G(x) \Leftrightarrow$ for all $x \in S, F(x) \leq G(x)$ and $F(x) \geq G(x) \Leftrightarrow F \sqsubseteq G$ and $G \sqsubseteq F$. ■

A ciset F on S is said to be an *c-empty ciset* if $F(x) = \langle 0, 0\rangle$ for all $x \in S$. We shall use $F_{\oslash}$ to denote the c-empty set. A ciset F on S is said to be the *c-universe ciset* if $F(x) = \langle 1, 1\rangle$ for all $x \in S$. We use $F_{\mathcal{U}(S)}$ to denote the c-universe ciset. Further, if there is no confusion, we shall abbreviate $F_{\mathcal{U}(S)}$ to $F_{\mathcal{U}}$. A ciset F on S is said to be a *c-singleton* if there exists a $y \in S$ such that $F(x) = \langle 0, 0\rangle$ for all $x \in S \neg \{y\}$ and $F(y) = \langle 0, 1\rangle$. We use I^y to denote the c-singleton. That is, $I^y(x) = \langle 0, 0\rangle$ for all $x \in S \neg \{y\}$ and $I^y(y) = \langle 0, 1\rangle$.

6.2 Ciset Operations

In this section we introduce five operations on cisets. They are c-union, c-intersection, c-difference, c-Cartesian product and c-negation. These operations allow us to construct new cisets from given cisets. We shall also study relationships among these operations.

Definition 6.2.1 *Let S be a set and let F, G be two cisets on S. The* c-union *of F and G, denoted by $F \sqcup G$ is a mapping from $S \to \mathfrak{C}$, defined by $(F \sqcup G)(x) = F(x) \sqcup G(x)$, for all $x \in S$.*

Clearly $F \sqcup G$ is the smallest ciset containing both F and G: In other words, if T is a ciset such that T contains both F and G, then T contains $F \sqcup G$.

Definition 6.2.2 *Let S be a set and let F, G be two cisets on S. The* c-intersection *of F and G, denoted by $F \sqcap G$ is a mapping from $S \to \mathfrak{C}$, defined by $(F \sqcap G)(x) = F(x) \sqcap G(x)$, for all $x \in S$.*

It may be noted that $F \sqcap G$ is the largest c-subset of both F and G. In other words, if T is a ciset such that T is a c-subset of both F and G, then $T \sqsubseteq F \sqcap G$. Two cisets F and G on S are said to be *disjoint* if $F \sqcap G = F_{\oslash}$, the empty set.

Definition 6.2.3 *Let S be a set and let F, G be two cisets on S. The* c-difference *of F and G, denoted by $F \neg G$ is a mapping from $S \to \mathfrak{C}$, defined by $(F \neg G)(x) = F(x) \neg G(x)$, for all $x \in S$.*

Definition 6.2.4 *Let S be a set and let F, G be two cisets on S. The* c-Cartesian product *of F and G, denoted by $F \times G$ is a mapping from $S \times S \to \mathfrak{C}$, defined by $(F \times G)(x, y) = F(x) \sqcap G(y)$, for all $(x, y) \in S \times S$.*

Definition 6.2.5 *Let S be a set and let F be a ciset on S. The* c-negation *of F, denoted by $\neg F$ is a mapping from $S \to \mathfrak{C}$, defined by $(\neg F)(x) = \neg F(x)$.*

Theorem 6.2.6 *Let F, G, H, J and K be any five cisets on a set S. Then we have the following:*

1. $F \sqcup F = F$,

2. $F \sqcap F = F$,

3. $F \sqcup G = G \sqcup F$,

4. $F \sqcap G = G \sqcap F$,

5. $F \sqcup F_{\oslash} = F$,

6. $F \sqcap F_{\mathcal{U}} = F$,

7. $F \sqcup F_{\mathcal{U}} = F_{\mathcal{U}}$,

8. $F \sqcap F_{\oslash} = F_{\oslash}$,

9. $F \sqcup F = F$,

10. $F \sqcap F = F$,

11. $F \sqcup (G \sqcup H) = (F \sqcup G) \sqcup H$,

12. $F \sqcap (G \sqcap H) = (F \sqcap G) \sqcap H$,

13. $F \sqcup (G \sqcap H) = (F \sqcup G) \sqcap (F \sqcup H)$,

14. $F \sqcap (G \sqcup F) = (F \sqcap G) \sqcup (F \sqcap H)$,

15. $\neg(F \sqcup G) = (\neg F) \sqcap (\neg G)$,

16. $\neg(F \sqcap G) = (\neg F) \sqcup (\neg G)$,

17. $(\neg(\neg F)) = F$,

18. *if* $F \leq G$ *then* $F \sqcup G = G$ *and* $F \sqcap G = F$,

19. *If* $F \sqsubseteq K, G \sqsubseteq J$ *then* $F \sqcup G \sqsubseteq K \sqcup J$,

20. *If* $F \sqsubseteq K, G \sqsubseteq J$ *then* $F \sqcap G \sqsubseteq K \sqcap J$.

Proof. Proof of each result follows from the corresponding result on confidence index. For example to prove (15), assume that $x \in S$. Then $(\neg(F \sqcup G))(x) = \neg((F \sqcup G)(x)) = \neg(F(x) \sqcup G(x)) = (\neg F(x)) \sqcap (\neg G(x)) = (\neg F)(x) \sqcap (\neg G)(x) = (\neg F) \sqcap (\neg G)(x)$. ■

It is worthwhile to note that the properties $F \sqcap (\neg F) = F_{\oslash}$ and $F \sqcup (\neg F) = F_{\mathcal{U}}$ do not hold in general. In logic, the former property is known as the *law of contradiction* and the latter property is called the *law of excluded middle.*

The reader may also note that (15) and (16) above prove *DeMorgan's Laws* do hold..

Proposition 6.2.7 *Let* $F, G,$ *and* H *be any three cisets on a set* S. *Then we have the following:*

1. $(F \sqcup G)\neg H = (F\neg H) \sqcup (G\neg H)$,

2. $F\neg(G \sqcap H) = (F\neg G) \sqcup (F\neg H)$,

3. $(F \sqcap G)\neg H = (F\neg H) \sqcap (G\neg H)$,

4. $F\neg(G \sqcup H) = (F\neg G) \sqcap (F\neg H)$,

Proof. Proof of each result follows from the corresponding result on confidence index. ■

A ciset can be considered a generalization of set. Let S be a set and A be any c-subset of S. Define a mapping $F_A : S \to \mathfrak{C}$ by $F_A(x) = \langle 0, 1\rangle$ if $x \in A$ and $F_A(x) = \langle 1, 0\rangle$ if $x \notin A$. Thus a ciset can be considered a generalization of set. Similarly, if $\mu : S \to [0, 1]$ is a fuzzy c-subset of S, define a mapping $F_\mu : S \to \mathfrak{C}$ by $F_\mu(x) = \langle 1 - \mu(x), \mu(x)\rangle$, for all $x \in S$. This convention and notation is followed for the rest of this book.

Theorem 6.2.8 *Let S be a set and $A, B \sqsubseteq S$. Then*

1. $F_A \sqcup F_B \sqsupseteq F_{A \sqcup B}$

2. $F_A \sqcap F_B \sqsubseteq F_{A \sqcap B}$

3. $F_A \neg F_B \sqsubseteq F_{A \neg B}$

4. $F_A \times F_B \sqsubseteq F_{A \times B}$

5. $(\neg F_A) = F_{A^c}$

Proof. Let $x \in S$. The proof can be summarized as follows. Proof of first three results follows from the following three tables. We distinguish four cases: (1) $x \in A, x \notin B$, (2) $x \notin A, x \in B$, (3) $x \in A, x \in B$ and (4) $x \notin A, x \notin B$.

As an illustration, we prove result (1) case (1) as follows: Assume that $x \in A, x \notin B$. Then $F_A(x) = \langle 0, 1\rangle, F_B(x) = \langle 1, 0\rangle$. Thus $(F_A \sqcup F_B)(x) = F_A(x) \sqcup F_B(x) = \langle 0 \vee 1, 1 \vee 0\rangle = \langle 1, 1\rangle$. Now $x \in A \sqcup B$. Therefore, $F_{A \sqcup B}(x) = \langle 0, 1\rangle$. Therefore, $F_A \sqcup F_B \sqsupseteq F_{A \sqcup B}$.

	F_A	F_B	$F_A \sqcup F_B$	$F_{A \sqcup B}$
$x \in A, x \notin B$	$\langle 0, 1\rangle$	$\langle 1, 0\rangle$	$\langle 1, 1\rangle$	$\langle 0, 1\rangle$
$x \notin A, x \in B$	$\langle 1, 0\rangle$	$\langle 0, 1\rangle$	$\langle 1, 1\rangle$	$\langle 0, 1\rangle$
$x \in A, x \in B$	$\langle 0, 1\rangle$	$\langle 0, 1\rangle$	$\langle 0, 1\rangle$	$\langle 0, 1\rangle$
$x \notin A, x \notin B$	$\langle 1, 0\rangle$	$\langle 1, 0\rangle$	$\langle 1, 0\rangle$	$\langle 1, 0\rangle$

	F_A	F_B	$F_A \sqcap F_B$	$F_{A \sqcap B}$
$x \in A, x \notin B$	$\langle 0, 1\rangle$	$\langle 1, 0\rangle$	$\langle 0, 0\rangle$	$\langle 1, 0\rangle$
$x \notin A, x \in B$	$\langle 1, 0\rangle$	$\langle 0, 1\rangle$	$\langle 0, 0\rangle$	$\langle 1, 0\rangle$
$x \in A, x \in B$	$\langle 0, 1\rangle$	$\langle 0, 1\rangle$	$\langle 0, 1\rangle$	$\langle 0, 1\rangle$
$x \notin A, x \notin B$	$\langle 1, 0\rangle$	$\langle 1, 0\rangle$	$\langle 1, 0\rangle$	$\langle 1, 0\rangle$

	F_A	F_B	$F_A \neg F_B$	$F_{A \neg B}$
$x \in A, x \notin B$	$\langle 0, 1\rangle$	$\langle 1, 0\rangle$	$\langle 0, 1\rangle$	$\langle 0, 1\rangle$
$x \notin A, x \in B$	$\langle 1, 0\rangle$	$\langle 0, 1\rangle$	$\langle 1, 0\rangle$	$\langle 1, 0\rangle$
$x \in A, x \in B$	$\langle 0, 1\rangle$	$\langle 0, 1\rangle$	$\langle 0, 0\rangle$	$\langle 1, 0\rangle$
$x \notin A, x \notin B$	$\langle 1, 0\rangle$	$\langle 1, 0\rangle$	$\langle 0, 0\rangle$	$\langle 1, 0\rangle$

Proof of result (4) follows from the following table. We distinguish four cases: (1) $x \in A, y \notin B$, (2) $x \notin A, y \in B$, (3) $x \in A, y \in B$ and (4) $x \notin A, y \notin B$.

	F_A	F_B	$F_A \times F_B$	$F_{A\times B}$
$x \in A, y \notin B$	$\langle 0,1\rangle$	$\langle 1,0\rangle$	$\langle 0,0\rangle$	$\langle 1,0\rangle$
$x \notin A, y \in B$	$\langle 1,0\rangle$	$\langle 0,1\rangle$	$\langle 0,0\rangle$	$\langle 1,0\rangle$
$x \in A, y \in B$	$\langle 0,1\rangle$	$\langle 0,1\rangle$	$\langle 0,1\rangle$	$\langle 0,1\rangle$
$x \notin A, y \notin B$	$\langle 1,0\rangle$	$\langle 1,0\rangle$	$\langle 1,0\rangle$	$\langle 1,0\rangle$

To prove the result (5), we distinguish two cases: (1) $x \in A$ and (2) $x \notin A$. The proof follows from following table.

	F_A	$(\neg F_A)$	F_{A^c}
$x \in A$	$\langle 0,1\rangle$	$\langle 1,0\rangle$	$\langle 1,0\rangle$
$x \notin A$	$\langle 1,0\rangle$	$\langle 0,1\rangle$	$\langle 0,1\rangle$

■

Theorem 6.2.9 *Let S be a set and let μ, σ be two fuzzy c-subsets on S.*

1. $F_\mu \sqcup F_\sigma \sqsupseteq F_{\mu\sqcup\sigma}$
2. $F_\mu \sqcap F_\sigma \sqsubseteq F_{\mu\sqcap\sigma}$
3. $F_\mu \times F_\sigma \sqsubseteq F_{\mu\times\sigma}$
4. $(\neg F_\mu) = F_{\mu^c}$

Proof. Let $x \in S$. Proof of first two results follows from the following two tables. We distinguish between two cases: (1) $\mu(x) \leq \sigma(x)$ and (2) $\mu(x) > \sigma(x)$. The first row of each of the following first two tables will correspond to the first case and the second row of each of the following first two tables will correspond to the second case respectively.

As an illustration, we prove result (1) case (1) as follows: Assume that $\mu(x) \leq \sigma(x)$. Then $F_\mu(x) = \langle 1-\mu(x), \mu(x)\rangle, F_\sigma(x) = \langle 1-\sigma(x), \sigma(x)\rangle$. Thus $(F_\mu \sqcup F_\sigma)(x) = F_\mu(x) \sqcup F_\sigma(x) = \langle (1-\mu(x)) \vee (1-\sigma(x)), \mu(x) \vee \sigma(x)\rangle = \langle 1-\mu(x), \sigma(x)\rangle$. Now $(\mu \sqcup \sigma)(x) = \mu(x) \sqcup \sigma(x) = \sigma(x)$. Hence $F_{\mu\sqcup\sigma}(x) = \langle 1-\sigma(x), \sigma(x)\rangle$. Therefore, $F_\mu \sqcup F_\sigma \sqsupseteq F_{\mu\sqcup\sigma}$.

F_μ	F_σ	$F_\mu \sqcup F_\sigma$	$F_{\mu\sqcup\sigma}$
$\langle 1-\mu(x), \mu(x)\rangle$	$\langle 1-\sigma(x), \sigma(x)\rangle$	$\langle 1-\mu(x), \sigma(x)\rangle$	$\langle 1-\sigma(x), \sigma(x)\rangle$
$\langle 1-\mu(x), \mu(x)\rangle$	$\langle 1-\sigma(x), \sigma(x)\rangle$	$\langle 1-\sigma(x), \mu(x)\rangle$	$\langle 1-\mu(x), \mu(x)\rangle$

F_μ	F_σ	$F_\mu \sqcap F_\sigma$	$F_{\mu\sqcap\sigma}$
$\langle 1-\mu(x), \mu(x)\rangle$	$\langle 1-\sigma(x), \sigma(x)\rangle$	$\langle 1-\sigma(x), \mu(x)\rangle$	$\langle 1-\mu(x), \mu(x)\rangle$
$\langle 1-\mu(x), \mu(x)\rangle$	$\langle 1-\sigma(x), \sigma(x)\rangle$	$\langle 1-\mu(x), \sigma(x)\rangle$	$\langle 1-\sigma(x), \sigma(x)\rangle$

Proof of result (3) follows from the following table. We distinguish between two cases: (1) $\mu(x) \leq \sigma(y)$ and (2) $\mu(x) > \sigma(y)$. The first row of each of the following table will correspond to the first case and the second row of the following table will correspond to the second case respectively.

F_μ	F_σ	$F_\mu \times F_\sigma$	$F_{\mu\times\sigma}$
$\langle 1-\mu(x), \mu(x)\rangle$	$\langle 1-\sigma(y), \sigma(y)\rangle$	$\langle 1-\sigma(y), \mu(x)\rangle$	$\langle 1-\mu(x), \mu(x)\rangle$
$\langle 1-\mu(x), \mu(x)\rangle$	$\langle 1-\sigma(y), \sigma(y)\rangle$	$\langle 1-\mu(x), \sigma(y)\rangle$	$\langle 1-\sigma(y), \sigma(y)\rangle$

The proof of (4) follows from following table.

F_μ	$(\neg F_\mu)$	F_{μ^c}
$\langle 1-\mu(x), \mu(x)\rangle$	$\langle \mu(x), 1-\mu(x)\rangle$	$\langle \mu(x), 1-\mu(x)\rangle$

■

Given a ciset $F : S \to \mathfrak{C}$, it is possible to produce various c-subsets and fuzzy c-subsets of S. In particular, we introduce the following. Let $\alpha = \langle s, t\rangle$ be a confidence index. Define $\alpha - c$*–cut set,* $F|_s^t$, by $F|_s^t = \{x \in S \mid u(F(x)) \geq t$ and $l(F(x)) \geq s\}$.

Proposition 6.2.10 *Let $a_1 = \langle s_1, t_1\rangle$ and $a_2 = \langle s_2, t_2\rangle$ be two confidence indexes such that $a_1 < a_2$ and let F be a ciset. Then $F|_{s_1}^{t_1} \sqsupseteq F|_{s_2}^{t_2}$.*

Proof. Let $x \in S$. Assume $x \in F|_{s_2}^{t_2}$. Then $u(F(x)) \geq t_2$ and $l(F(x)) \geq s_2$. Therefore, $u(F(x)) \geq t_1$ and $l(F(x)) \geq s_1$. Hence $x \in F|_{s_1}^{t_1}$. ■

Proposition 6.2.11 *Let $a = \langle s, t\rangle$ be two confidence index and let F, G be two cisets. Then $(F \sqcup G)|_s^t \sqsupseteq F|_s^t \sqcup G|_s^t$ and $(F \sqcap G)|_s^t = F|_s^t \sqcap G|_s^t$.*

Proof. Let $x \in S$. Now, $x \in (F \sqcap G)|_s^t \Leftrightarrow u((F \sqcap G)(x)) \geq t$ and $l((F \sqcap G)(x)) \geq s \Leftrightarrow u(F(x)) \geq t$ and $l(F(x)) \geq s$; and $u(G(x)) \geq t$ and $l(G(x)) \geq s \Leftrightarrow x \in F|_s^t \sqcap G|_s^t$. ■

If $s = 0$, then the symbol $F|^t$ is used in place of $F|_0^t$ and similarly, if $t = 0$, then the symbol $F|_s$ is used in place of $F|_s^0$. Further, $F|^0$ is used instead of $F|_0^0$. The *upper fuzzy set* corresponding to the ciset F is defined as $\overline{F} = \{(x, u(F(x))) \mid x \in S\}$. In other words, $\overline{F}$ is a mapping from S into $[0, 1]$ such that $\overline{F}(x) = u(F(x))$. Similarly, the *lower fuzzy set* corresponding to the ciset F is defined as $\underline{F} = \{(x, l(F(x))) \mid x \in S\}$. In other words, $\underline{F}$ is a mapping from S into $[0, 1]$ such that $\underline{F}(x) = l(F(x))$.

Theorem 6.2.12 *Let S be a set and $A, B \sqsubseteq S$. Then*

1. $(F_A \sqcup F_B)|^1 = (F_A)|^1 \sqcup (F_B)|^1 = (F_{A\sqcup B})|^1$
2. $(F_A \sqcap F_B)|^1 = (F_A)|^1 \sqcap (F_B)|^1 = (F_{A\sqcap B})|^1$
3. $(F_A \neg F_B)|^1 = (F_A)|^1 \neg (F_B)|^1 = (F_{A\neg B})|^1$
4. $(F_A \times F_B)|^1 = (F_A)|^1 \times (F_B)|^1 = (F_{A\times B})|^1$
5. $(\neg F_A)|^1 = (F_{A^c})|^1$

Proof. Let $x \in S$. The proof can be summarized as follows. Proof of first three results follows from the following three tables. We distinguish four cases: (1) $x \in A, x \notin B$, (2) $x \notin A, x \in B$, (3) $x \in A, x \in B$ and (4) $x \notin A, x \notin B$.

As an illustration, we prove result (1) case (1) as follows: Assume that $x \in A, x \notin B$. Then $F_A(x) = \langle 0,1\rangle, F_B(x) = \langle 1,0\rangle$. Consequently, $x \in (F_A)|^1$, $x \notin (F_B)|^1$ and thus $x \in (F_A)|^1 \sqcup (F_B)|^1$. Also, $(F_A \sqcup F_B)(x) = F_A(x) \sqcup F_B(x) = \langle 0 \vee 1, 1 \vee 0\rangle = \langle 1,1\rangle$ and thus $x \in (F_A \sqcup F_B)|^1$. Now $x \in A \sqcup B$. Therefore, $F_{A\sqcup B}(x) = \langle 0,1\rangle$ and hence $x \in (F_{A\sqcup B})|^1$. Thus $(F_A \sqcup F_B)|^1 = (F_A)|^1 \sqcup (F_B)|^1 = (F_{A\sqcup B})|^1$.

	$x \in (F_A)\|^1$	$x \in (F_B)\|^1$	$x \in (F_A \sqcup F_B)\|^1$	$x \in (F_{A\sqcup B})\|^1$
(1)	T	F	T	T
(2)	F	T	T	T
(3)	T	T	T	T
(4)	F	F	F	F

	$x \in (F_A)\|^1$	$x \in (F_B)\|^1$	$x \in (F_A \sqcap F_B)\|^1$	$x \in (F_{A\sqcap B})\|^1$
(1)	T	F	F	F
(2)	F	T	F	F
(3)	T	T	T	T
(4)	F	F	F	F

	$x \in (F_A \neg F_B)\|^1$	$x \in (F_{A\neg B})\|^1$
$x \in A, x \notin B$	T	T
$x \notin A, x \in B$	F	F
$x \in A, x \in B$	F	F
$x \notin A, x \notin B$	F	F

Proof of result (4) follows from the following table. We distinguish four cases: (1) $x \in A, y \notin B$, (2) $x \notin A, y \in B$, (3) $x \in A, y \in B$ and (4) $x \notin A, y \notin B$.

	$x \in (F_A \times F_B)\|^1$	$x \in (F_{A\times B})\|^1$
$x \in A, y \notin B$	F	F
$x \notin A, y \in B$	F	F
$x \in A, y \in B$	T	T
$x \notin A, y \notin B$	F	F

To prove the result (5), we distinguish two cases: (1) $x \in A$ and (2) $x \notin A$. The proof follows from following table.

	$x \in (\neg F_A)\|^1$	$x \in (F_{A^c})\|^1$
$x \in A$	F	F
$x \notin A$	T	T

■

Theorem 6.2.13 *Let S be a set and let μ, σ be two fuzzy c-subsets on S.*

1. $\overline{F_\mu \sqcup F_\sigma} = \overline{F_\mu} \sqcup \overline{F_\sigma} = \overline{F_{\mu\sqcup\sigma}}$
2. $\overline{F_\mu \sqcap F_\sigma} = \overline{F_\mu} \sqcap \overline{F_\sigma} = \overline{F_{\mu\sqcap\sigma}}$
3. $\overline{F_\mu \times F_\sigma} = \overline{F_\mu} \times \overline{F_\sigma} = \overline{F_{\mu\times\sigma}}$

4. $\overline{\neg F_\mu} = \overline{F_{\mu^c}}$

Proof. Let $x \in S$. The proof is summarized in the following tables. Proof of first two results follows from the following two tables. We distinguish between two cases: (1) $\mu(x) \leq \sigma(x)$ and (2) $\mu(x) > \sigma(x)$.

As an illustration, we prove result (1) case (1) as follows: Assume that $\mu(x) \leq \sigma(x)$. Then $F_\mu(x) = \langle 1-\mu(x), \mu(x)\rangle$, $F_\sigma(x) = \langle 1-\sigma(x), \sigma(x)\rangle$. Hence $(\overline{F_\mu} \sqcup \overline{F_\sigma})(x) = \overline{F_\mu}(x) \sqcup \overline{F_\sigma}(x) = \mu(x) \vee \sigma(x) = \sigma(x)$. Now $(F_\mu \sqcup F_\sigma)(x) = F_\mu(x) \sqcup F_\sigma(x) = \langle (1-\mu(x)) \vee (1-\sigma(x)), \mu(x) \vee \sigma(x)\rangle = \langle 1-\mu(x), \sigma(x)\rangle$ and hence $\overline{F_\mu \sqcup F_\sigma}(x) = \sigma(x)$. Note also that $\overline{(\mu \sqcup \sigma)(x)} = \mu(x) \sqcup \sigma(x) = \sigma(x)$. Hence $F_{\mu\sqcup\sigma}(x) = \langle 1-\sigma(x), \sigma(x)\rangle$ and thus $\overline{F_{\mu\sqcup\sigma}}(x) = \sigma(x)$. Therefore, $\overline{F_\mu \sqcup F_\sigma} = \overline{F_\mu} \sqcup \overline{F_\sigma} = \overline{F_{\mu\sqcup\sigma}}$.

	$u(F_\mu(x))$	$u(F_\sigma(x))$	$u((F_\mu \sqcup F_\sigma)(x))$	$u(F_{\mu\sqcup\sigma}(x))$
$\mu(x) \leq \sigma(x)$	$\mu(x)$	$\sigma(x)$	$\sigma(x)$	$\sigma(x)$
$\mu(x) > \sigma(x)$	$\mu(x)$	$\sigma(x)$	$\mu(x)$	$\mu(x)$

	$u((F_\mu \sqcap F_\sigma)(x))$	$u(F_{\mu\sqcap\sigma}(x))$
$\mu(x) \leq \sigma(x)$	$\mu(x)$	$\mu(x)$
$\mu(x) > \sigma(x)$	$\sigma(x)$	$\sigma(x)$

Proof of result (3) follows from the following table. We distinguish between two cases: (1) $\mu(x) \leq \sigma(y)$ and (2) $\mu(x) > \sigma(y)$.

	$u((F_\mu \times F_\sigma)(x))$	$u(F_{\mu\times\sigma}(x))$
$\mu(x) \leq \sigma(y)$	$\mu(x)$	$\mu(x)$
$\mu(x) > \sigma(y)$	$\sigma(y)$	$\sigma(y)$

The proof of (4) follows from following table.

$u(\neg F_\mu(x))$	$u(F_{\mu^c}(x))$
$1 - \mu(x)$	$1 - \mu(x)$

■

Theorem 6.2.14 *Let S be a set and $A, B \sqsubseteq S$. Then the following diagrams commute.*

$$\begin{array}{ccc} (A,B) & \xrightarrow{F} & (F_A, F_B) \\ \downarrow \sqcup & & \downarrow \sqcup \\ A \sqcup B & \xleftarrow{\langle 0,1\rangle\text{-c-cut}} & F_A \sqcup F_B \end{array}$$

$$\begin{array}{ccc} (A,B) & \xrightarrow{F} & (F_A, F_B) \\ \downarrow \sqcap & & \downarrow \sqcap \\ A \sqcap B & \xleftarrow{\langle 0,1\rangle\text{-c-cut}} & F_A \sqcap F_B \end{array}$$

$$\begin{array}{ccc} (A,B) & \xrightarrow{F} & (F_A, F_B) \\ \Big\downarrow \neg & & \Big\downarrow \neg \\ A \neg B & \xleftarrow{\langle 0,1 \rangle\text{-c-cut}} & F_A \neg F_B \end{array}$$

$$\begin{array}{ccc} (A,B) & \xrightarrow{F} & (F_A, F_B) \\ \Big\downarrow \times & & \Big\downarrow \times \\ A \times B & \xleftarrow{\langle 0,1 \rangle\text{-c-cut}} & F_A \times F_B \end{array}$$

$$\begin{array}{ccc} A & \xrightarrow{F} & F_A \\ \Big\downarrow c & & \Big\downarrow \neg \\ A^c & \xleftarrow{\langle 0,1 \rangle\text{-c-cut}} & \neg F_A \end{array}$$

Proof. Result follows from Theorem 6.2.12 ■

The Theorem 6.2.14 is a very powerful result. It states that bgiven two sets A and B, the five operations c-union, c-intersection, c-difference, Cartesian product and c-negation, can either be computed as such or A and B can be converted into equivalent cisets and the corresponding operations can be carried out as cisets and converted back to sets without any loss of information. Since this result holds for all five operations, any operation defined using these five operations shall also inherit this property. This result clearly demonstrates the fact that the ciset is a generilzation of the set.

Theorem 6.2.15 *Let S be a set and μ, σ be two fuzzy sets on S. Then the following diagrams commute.*

$$\begin{array}{ccc} (\mu,\sigma) & \xrightarrow{F} & (F_\mu, F_\sigma) \\ \Big\downarrow \sqcup & & \Big\downarrow \sqcup \\ \mu \sqcup \sigma & \xleftarrow{\overline{F}} & F_\mu \sqcup F_\sigma \end{array}$$

$$\begin{array}{ccc} (\mu,\sigma) & \xrightarrow{F} & (F_\mu, F_\sigma) \\ \Big\downarrow \sqcap & & \Big\downarrow \sqcap \\ A \sqcap B & \xleftarrow{\overline{F}} & F_\mu \sqcap F_\sigma \end{array}$$

$$\begin{array}{ccc} (\mu,\sigma) & \xrightarrow{F} & (F_\mu, F_\sigma) \\ \downarrow{\times} & & \downarrow{\times} \\ A \times B & \xleftarrow{\overline{F}} & F_\mu \times F_\sigma \end{array}$$

$$\begin{array}{ccc} \mu & \xrightarrow{F} & F_\mu \\ \downarrow{c} & & \downarrow{\neg} \\ \mu^c & \xleftarrow{\overline{F}} & \neg F_\mu \end{array}$$

Proof. Result follows from Theorem6.2.13 ■

The Theorem 6.2.15 is the fuzzy equivalent of the Theorem 6.2.14. In simple terms, it states that given two fuzzy c-subsets sets μ and σ, the four operations c-union, c-intersection, Cartesian product and c-negation can either be computed as fuzzy c-subsets or μ and σ can be converted into equivalent cisets and the corresponding operations can be carried out as cisets and then converted back to fuzzy sets without any loss of information. Since this result holds for all four operations, any operation defined using these operations shall also inherit this property.This result clearly demonstrates the fact that the ciset is a generilzation of the fuzzy set.

In fact, from the above discussion, we have the following two results as well.

Theorem 6.2.16 *Let S be a set and $A, B \sqsubseteq S$. Then the following diagrams commute.*

$$\begin{array}{ccc} (A,B) & \xrightarrow{F} & (F_A, F_B) \\ \downarrow{\sqcup} & & \downarrow{\sqcup} \\ A \sqcup B & \xrightarrow{F} & F_A \sqcup F_B \end{array}$$

$$\begin{array}{ccc} (A,B) & \xrightarrow{F} & (F_A, F_B) \\ \downarrow{\sqcap} & & \downarrow{\sqcap} \\ A \sqcap B & \xrightarrow{F} & F_A \sqcap F_B \end{array}$$

$$\begin{array}{ccc} (A,B) & \xrightarrow{F} & (F_A, F_B) \\ \downarrow{\neg} & & \downarrow{\neg} \\ A \neg B & \xrightarrow{F} & F_A \neg F_B \end{array}$$

$$\begin{array}{ccc} (A,B) & \xrightarrow{F} & (F_A,F_B) \\ \downarrow \times & & \downarrow \times \\ A\times B & \xrightarrow{F} & F_A\times F_B \end{array}$$

$$\begin{array}{ccc} A & \xrightarrow{F} & F_A \\ \downarrow c & & \downarrow \neg \\ A^c & \xrightarrow{F} & \neg F_A \end{array}$$

■

Theorem 6.2.17 *Let S be a set and μ, σ be two fuzzy sets on S. Then the following diagrams commute.*

$$\begin{array}{ccc} (\mu,\sigma) & \xrightarrow{F} & (F_\mu,F_\sigma) \\ \downarrow \sqcup & & \downarrow \sqcup \\ \mu\sqcup\sigma & \xrightarrow{F} & F_\mu\sqcup F_\sigma \end{array}$$

$$\begin{array}{ccc} (\mu,\sigma) & \xrightarrow{\overline{F}} & (F_\mu,F_\sigma) \\ \downarrow \sqcap & & \downarrow \sqcap \\ A\sqcap B & \xrightarrow{F} & F_\mu\sqcap F_\sigma \end{array}$$

$$\begin{array}{ccc} (\mu,\sigma) & \xrightarrow{F} & (F_\mu,F_\sigma) \\ \downarrow \times & & \downarrow \times \\ A\times B & \xrightarrow{F} & F_\mu\times F_\sigma \end{array}$$

$$\begin{array}{ccc} \mu & \xrightarrow{F} & F_\mu \\ \downarrow c & & \downarrow \neg \\ \mu^c & \xrightarrow{F} & \neg F_\mu \end{array}$$

■

6.3 C-relations

Let S and T be two sets and let G and H be two cisets on S and T respectively. Then a c-subset $K \sqsubseteq G \times H$ is called a *c-relation* from G to H. In other words, K is a ciset on $S \times T$ such that $K(x, y) \leq G(x) \sqcap H(y)$, for all $(x, y) \in S \times T$. Thus the confidence indexes of a pair of elements never exceed the confidence index of either of the elements themselves. If we associate the elements as computers and pairs as the connecting communication links between the computers, the above restriction amounts to requiring that the confidence index of a communication link can never exceed the strengths of its connecting computers. The restriction $K(x, y) \sqsubseteq G(x) \sqcap H(y)$, for all $(x, y) \in S \times T$ allows $K|_s^t$ to be a c-relation from $G|_s^t$ into $H|_s^t$ for all $\langle s, t\rangle \in \mathfrak{C}$.

There are three special cases of c-relations.

case 1: $S = T$ and $G = H$. In this case K is said to be a c-relation on G. Note that K is a c-subset of $G \times G$. Thus $K(x, y) \sqsubseteq G(x) \sqcap G(y)$, for all $x, y \in S$.

case 2: $G(x) = \langle 0, 1\rangle$, for all $x \in S$ and $H(y) = \langle 0, 1\rangle$, for all $y \in T$. In this case, whenever there is no confusion, we may say K is a c-relation from S into T.

case 3: $S = T, G(x) = H(x) = \langle 0, 1\rangle$, for all $x \in S$. In this case, whenever there is no confusion, we may say K is a c-relation on S.

Example 6.3.1 *Let* $S = \{u, x, y, z\}$. *Define* $G(u) = \langle 0.8, 0.3\rangle$, $G(x) = \langle 0.3, 0.7\rangle, G(y) = \langle 0.2, 0.4\rangle$, $G(z) = \langle 0, 0.5\rangle$. *Clearly* G *is a ciset on* S. *Let* K *be ciset on* $S \times S$ *defined as follows:*

	u	x	y	z
u	$\langle 0.5, 0.2\rangle$	$\langle 0.2, 0.3\rangle$	$\langle 0.1, 0.2\rangle$	$\langle 0, 0.3\rangle$
x	$\langle 0.2, 0.3\rangle$	$\langle 0.1, 0.7\rangle$	$\langle 0.2, 0.2\rangle$	$\langle 0, 0.5\rangle$
y	$\langle 0, 0\rangle$	$\langle 0.5, 0.4\rangle$	$\langle 0.2, 0.3\rangle$	$\langle 0, 0.4\rangle$
z	$\langle 0, 0\rangle$	$\langle 0, 0\rangle$	$\langle 0, 0.2\rangle$	$\langle 0, 0.1\rangle$

Since $K(y, x) = \langle 0.5, 0.4\rangle \not\leq \langle 0.3, 0.4\rangle = G(x) \sqcap H(y)$, K is not a c-relation on G. However, if we redefine $K(y, x) = \langle 0.3, 0.4\rangle$ then K is a c-relation on G.

Let K be a c-relation on S. Then K is called the strongest c-relation on G if for all c-relations J on G, $J \leq K$, treating both J and K as c-subsets of $G \times G$. It is quite clear that K is the strongest c-relation on G, if and only if $K(x.y) = G(x) \sqcap K(y)$, for all $x, y \in S$.

Example 6.3.2 *Let* $S = \{u, x, y, z\}$. *Define* $G(u) = \langle 0.9, 0.7\rangle$, $G(x) = \langle 0.5, 0.9\rangle, G(y) = \langle 0.6, 0.8\rangle$, $G(z) = \langle 0.4, 0.6\rangle$. *Thus,* G *is a ciset on* S. *Let*

K be ciset on $S \times S$ be as follows:

	u	x	y	z
u	$\langle 0.9, 0.7\rangle$	$\langle 0.5, 0.7\rangle$	$\langle 0.6, 0.7\rangle$	$\langle 0.4, 0.6\rangle$
x	$\langle 0.5, 0.7\rangle$	$\langle 0.5, 0.9\rangle$	$\langle 0.5, 0.8\rangle$	$\langle 0.4, 0.6\rangle$
y	$\langle 0.6, 0.7\rangle$	$\langle 0.5, 0.8\rangle$	$\langle 0.6, 0.8\rangle$	$\langle 0.4, 0.6\rangle$
z	$\langle 0.4, 0.6\rangle$	$\langle 0.4, 0.6\rangle$	$\langle 0.4, 0.6\rangle$	$\langle 0.4, 0.6\rangle$

It is easy to observe that K is the strongest c-relation on G.

The converse problem may also arise in practice. That is, we may know the confidence index of the pairs and we want to compute the minimum confidence index required for the elements themselves. For a given ciset K of $S \times S$, the weakest ciset G on S such that K is a c-relation on G is obtained by $G(x) = \sqcup\{K(x,y) \sqcup K(y,x) \mid y \in S\}$ for all $x \in S$.

Example 6.3.3 *Let $S = \{u, x, y, z\}$. Let K be ciset on $S \times S$ defined as follows:*

	u	x	y	z
u	$\langle 0.5, 0.2\rangle$	$\langle 0.2, 0.3\rangle$	$\langle 0.1, 0.2\rangle$	$\langle 0, 0.3\rangle$
x	$\langle 0.2, 0.3\rangle$	$\langle 0.1, 0.7\rangle$	$\langle 0.2, 0.2\rangle$	$\langle 0, 0.5\rangle$
y	$\langle 0, 0\rangle$	$\langle 0.2, 0.4\rangle$	$\langle 0.2, 0.3\rangle$	$\langle 0, 0.4\rangle$
z	$\langle 0.5, 0\rangle$	$\langle 0, 0\rangle$	$\langle 0, 0.2\rangle$	$\langle 0, 0.1\rangle$

Define $G : S \to \mathfrak{C}$ as below.

$S(u) = \langle 0.5, 0.2\rangle \sqcup \langle 0.2, 0.3\rangle \sqcup \langle 0.1, 0.2\rangle \sqcup \langle 0, 0.3\rangle \sqcup \langle 0.2, 0.3\rangle \sqcup \langle 0, 0\rangle \sqcup \langle 0.5, 0\rangle$,

$S(x) = \langle 0.2, 0.3\rangle \sqcup \langle 0.1, 0.7\rangle \sqcup \langle 0.2, 0.2\rangle \sqcup \langle 0, 0.5\rangle \sqcup \langle 0.2, 0.3\rangle \sqcup \langle 0.2, 0.4\rangle \sqcup \langle 0, 0\rangle$,

$S(y) = \langle 0, 0\rangle \sqcup \langle 0.2, 0.4\rangle \sqcup \langle 0.2, 0.3\rangle \sqcup \langle 0, 0.4\rangle \sqcup \langle 0.1, 0.2\rangle \sqcup \langle 0.2, 0.2\rangle \sqcup \langle 0, 0.2\rangle$,

$S(z) = \langle 0.5, 0\rangle \sqcup \langle 0, 0\rangle \sqcup \langle 0, 0.2\rangle \sqcup \langle 0, 0.1\rangle \sqcup \langle 0, 0.3\rangle \sqcup \langle 0, 0.5\rangle \sqcup \langle 0, 0.4\rangle$.

Thus $S(u) = \langle 0.5, 0.3\rangle$, $S(x) = \langle 0.2, 0.7\rangle$, $S(y) = \langle 0.2, 0.4\rangle$, $S(z) = \langle 0.5, 0.5\rangle$. Clearly K is c-relation on G and G is the weakest ciset on S such that K is a c-relation on G.

Given a c-relation K on a ciset G, it is possible to take (s,t)–cuts of both K and G to obtain a c-relation $K|_s^t$ on $G|_s^t$.

Example 6.3.4 *Let $S = \{u, x, y, z\}$. Define a ciset G on S as follows: $G(u) = \langle 0.8, 0.3\rangle$, $G(x) = \langle 0.3, 0.7\rangle, G(y) = \langle 0.2, 0.4\rangle$, $G(z) = \langle 0.6, 0.5\rangle$. Let K be ciset on $S \times S$ defined as follows:*

	u	x	y	z
u	$\langle 0.5, 0.2\rangle$	$\langle 0.2, 0.3\rangle$	$\langle 0.1, 0.2\rangle$	$\langle 0, 0.3\rangle$
x	$\langle 0.2, 0.3\rangle$	$\langle 0.5, 0.7\rangle$	$\langle 0.2, 0.6\rangle$	$\langle 0.4, 0.5\rangle$
y	$\langle 0, 0\rangle$	$\langle 0.2, 0.4\rangle$	$\langle 0.8, 0.7\rangle$	$\langle 0.3, 0.4\rangle$
z	$\langle 0, 0\rangle$	$\langle 0, 0\rangle$	$\langle 0.3, 0.6\rangle$	$\langle 0, 0.1\rangle$

Choose $s = 0.4$ and $t = 0.2$. Then
$G_{0.2}^{0.4} = \{x, y, z\}$ and $K|_{0.2}^{0.4} = \{(x,x), (x,y), (x,z), (y,x), (y,y), (y,z), (z,y)\}$.

Operations on c-relations

Let S be a set and G be a ciset on S. Given two c-relations J and K on G, J and K are cisets on $S \times S$. Therefore $J = K, J < K, J \leq K, J > K, J \geq K, J \sqcup K, J \sqcap K, \neg K$,and $J \neg K$ are defined. In particular, $J \sqcup K, J \sqcap K, \neg K, J \neg K$ are cisets on $S \times S$ and thus they are c-relations on G. For example, $J \sqcup K$ and $J \sqcap K$ are defined as follows: $(J \sqcup K)(x, y) = J(x, y) \sqcup K(x, y), (J \sqcap K)(x, y) = J(x, y) \sqcap K(x, y)$. Note that in particular, Theorem 6.2.6 hold.

We now proceed to introduce a new binary operation.

Definition 6.3.5 *Let S, T, U be sets and F, G, H be cisets on S, T and U respectively. Let K be c-relation from F to G and J be c-relation from G to H. Define $K \circ J : S \times U \to \mathfrak{C}$ by*

$$K \circ J(x, z) = \sqcup\{K(x, y) \sqcap J(y, z) \mid y \in T\}$$

for all $x \in S, z \in U$. The c-relation $K \circ J$ is called the *composition* of K with J.

Note that $K \circ J$ is a c-relation from ciset F to ciset H. A closer look at the definition of the composition operation reveals that $K \circ J$ can be computed similar to matrix multiplication, where the arithmetic operation addition is replaced by confidence index operation c-union and arithmetic operation multiplication is replaced by confidence index operation c-intersection.

Proposition 6.3.6 *The composition operation is associative.*

Note that in the special case, $S = T$ and $F = G, K \circ K$ is defined. Since composition is associative, we use the notation K^2 to denote $K \circ K$ and K^n to denote $K^{n-1} \circ K, n > 2$. Define $K^\infty(x, y) = \sqcup\{K^n(x, y) \mid n > 1\} \sqcup K(x, y)$. It may be recalled that we already use K^1 to denote $\langle 0, 1\rangle$–cut of K. Therefore, we shall avoid the notation K^1 to denote K. Further define K^0 by $K^0(x, y) = \langle 0, 0\rangle$ and $K^0(x, x) = F(x)$ otherwise.

Definition 6.3.7 *Let S, T be sets and F, G be cisets on S, T respectively. Let K be c-relation from F to G. Define $K^{\neg 1} : T \times S \to \mathfrak{C}$ by*

$$K^{\neg 1}(x, y) = K(y, x)$$

for all $x \in T, y \in S$. The c-relation $K^{\neg 1}$is called the *inverse* of K.

We have the following result, proof of which is quite straight forward and hence we omit it.

Proposition 6.3.8 *Let S, T, U be sets and F, G, H be cisets on S, T, U respectively. Let K, K' be c-relations from F to G and J, J' be c-relation from G to H. Further, let $a = \langle s, t\rangle \in \mathfrak{C}$. Then*

1. $(K \circ J)|_s^t \sqsupseteq K|_s^t \circ J|_s^t$ and if S, T and U are finite, $(K \circ J)|_s^t = K|_s^t \circ J|_s^t$
2. If $K \sqsubseteq K'$ and $J \sqsubseteq J'$ then $K \circ J \sqsubseteq K' \circ J'$

Equivalence c-relations

Throughout this section, let S be a set and F be a ciset on S. Let K and J be two c-relations on F. It is quite natural to represent a c-relation in the form of a matrix. We now use the matrix representation of a c-relation to explain the properties of a c-relation. In particular, we shall use the term diagonal to represent the principal diagonal of the matrix.

We call K *reflexive* on F, if $K(x,x) = F(x)$ for all $x \in S$. If K is reflexive on F then $K(x,y) \leq F(x) \sqcap F(y) \leq F(x) = K(x,x)$ and it follows that any diagonal element is larger than or equal to any element in its row. Similarly, any diagonal element is larger than or equal to any element in its column. Conversely, given a c-relation K on F such that any diagonal element is larger than or equal to any element in its row and column, define a ciset G on S as $G(x) = K(x,x)$, for all $x \in S$. Then G is the weakest ciset on S such that K is a c-relation on G. Further, K is reflexive on G.

Reflexive c-relations have some interesting algebraic properties.

Theorem 6.3.9 *Let S be a set and F be a ciset on S and let K and J be two c-relations on F. Then the following properties hold.*

1. If K is reflexive, $J \sqsubseteq J \circ K$ and $J \sqsubseteq K \circ J$,
2. If K is reflexive, $K^0(x,x) = K(x,x) = K^2(x,x) = \ldots = K^\infty(x,x) = F(x), \forall x \in S$,
3. If K is reflexive, $K^0 \sqsubseteq K \sqsubseteq K^2 \sqsubseteq \ldots \sqsubseteq K^\infty$,
4. If K and J are reflexive, so is $J \circ K$ and $K \circ J$,
5. If K is reflexive, then $K|_s^t$ is a reflexive c-relation on $F|_s^t$ for all confidence indexes $\langle s,t \rangle$.

Proof. Let $x, z \in S$.

1. $(K \circ J)(x,z) = \sqcup\{K(x,y) \sqcap J(y,z) \mid y \in S\} \geq K(x,x) \sqcap J(x,z) = F(x) \sqcap J(x,z)$. Recall that $J(x,z) \leq F(x) \sqcap F(z)$. Therefore, $F(x) \sqcap J(x,z) = J(x,z)$. Thus $J \sqsubseteq J \circ K$. The result, $J \sqsubseteq K \circ J$ follows similarly.
2. Note that $K(x,x) = F(x), \forall x \in S$. Assume that $K^n(x,x) = F(x)$, $\forall x \in S$. Now for all $x \in S$, $K^{n+1}(x,x) = \sqcup\{K(x,y) \sqcap K^n(y,x) \mid y \in S\} \leq \sqcup\{F(x) \sqcap F(x) \mid y \in S\} = F(x)$ and $K^{n+1}(x,x) = \sqcup\{K(x,y) \sqcap K^n(y,x) \mid y \in S\} \geq K(x,x) \sqcap K^n(x,x) = F(x)$. Hence the result.

3. Choose J to be the same as K in (1). Thus we get $K \sqsubseteq K^2$. By (2), K^2 is reflexive. Now apply (1) by choosing J as K^2. Thus $K^2 \sqsubseteq K^3$. Repeating this process, the result follows.

4. $(K \circ J)(x,x) = \sqcup\{K(x,y) \sqcap J(y,x) \mid y \in S\} \leq \sqcup\{F(x) \sqcap F(x) \mid y \in S\} = F(x)$ and $(K \circ J)(x,x) = \sqcup\{K(x,y) \sqcap J(y,x) \mid y \in S\} \geq K(x,x) \sqcap J(x,x) = F(x) \sqcap F(x) = F(x)$. The proof that $K \circ J$ is reflexive is similar.

5. If $x \in F|_s^t$, then $\langle s,t \rangle \leq F(x) = K(x,x)$. Therefore $(x,x) \in K|_s^t$.

■

We call K *symmetric* if $K(x,y) = K(y,x)$, for all $x,y \in S$. In other words, K is symmetric if the matrix representation of K is symmetric.

Theorem 6.3.10 *Let S be a set and F be a ciset on S and let K and J be two c-relations on F. Then the following properties hold.*

1. If K and J are symmetric then $J \circ K$ is symmetric if and only if $J \circ K = K \circ J$,

2. If K is symmetric, then so is every power of K,

3. If K is symmetric, then $K|_s^t$ is a symmetric c-relation on $F|_s^t$ for all confidence indexes $\langle s,t \rangle$.

Proof. Let $x, z \in S$.

1. $(K \circ J)(x,z) = (K \circ J)(z,x)$
   $\Leftrightarrow \sqcup\{K(x,y) \sqcap J(y,z) \mid y \in S\} = \sqcup\{K(z,y) \sqcap J(y,x) \mid y \in S\}$
   $\Leftrightarrow \sqcup\{K(x,y) \sqcap J(y,z) \mid y \in S\} = \sqcup\{J(x,y) \sqcap K(y,z) \mid y \in S\}$
   $\Leftrightarrow K \circ J = J \circ K$

2. Assume that K^n is symmetric for $n \in \mathbb{N}$. Then $K^{n+1}(x,z)$
   $= \sqcup\{K(x,y) \sqcap K^n(y,z) \mid y \in S\}$
   $= \sqcup\{K(y,x) \sqcap K^n(z,y) \mid y \in S\}$
   $= \sqcup\{K^n(z,y) \sqcap K(y,x) \mid y \in S\}$
   $= K^{n+1}(z,x)$. Hence the result.

3. If $(x,z) \in K|_s^t$ then $\langle s,t \rangle \leq K(x,z) = K(z,x)$. Therefore $(z,x) \in K|_s^t$.

■

We call K transitive if $K^2 \sqsubseteq K$. It follows that K^∞ is always transitive for any c-relation K.

Theorem 6.3.11 *Let S be set and F be cisets on S and let K, J and L be three c-relations on F. Then the following properties hold.*

1. If K is transitive and $J \sqsubseteq K, L \sqsubseteq K$ then $J \circ L \sqsubseteq K$,
2. If K is transitive, then so is every power of K,
3. If K is transitive, J is reflexive and $J \sqsubseteq K$ then $K \circ J = J \circ K = K$,
4. If K is reflexive and transitive then $K^0 \sqsubseteq K = K^2 = \ldots = K^\infty$,
5. If K and J are transitive and $J \circ K = K \circ J$, then $K \circ J$ is transitive,
6. If K is symmetric and transitive then $K(x,y) \leq K(x,x)$ and $K(y,x) \leq K(x,x)$, for all $x, y \in S$,
7. If K is transitive, then $K|_s^t$ is a reflexive c-relation on F_s^t for all confidence indexes $\langle s,t \rangle$.

Proof. Let $x, z, w \in S$.

1. $(J \circ L)(x,z) = \sqcup\{J(x,y) \sqcap L(y,z) \mid y \in S\} \leq \sqcup\{K(x,y) \sqcap K(y,z) \mid y \in S\} = K^2(x,z) \leq K(x,z)$.
2. Assume that K^n is transitive. Then $K^n \circ K^n \sqsubseteq K^n$. Now $K^{n+1} \circ K^{n+1} = K^{2n} \circ K^2 \sqsubseteq K^n \circ K = K^{n+1}$.
3. By (1), $K \circ J \sqsubseteq K$. Note that $(K \circ J)(x,z) = \sqcup\{K(x,y) \sqcap J(y,z) \mid y \in S\} \geq K(x,z) \sqcap J(z,z) = K(x,z) \sqcap F(z) = K(x,z)$. Hence $K \circ J = K$. Similarly, $J \circ K = K$.
4. By (3), $K \circ K = K$. Thus $K^2 = K$. Assume that $K^n = K$. Since K^n is transitive and K is reflexive, by (3), $K^n \circ K = K$. Thus $K^{n+1} = K$.
5. $(K \circ J) \circ (K \circ J) = K \circ (J \circ K) \circ J = K \circ (K \circ J) \circ J \sqsubseteq K \circ J$. Thus $K \circ J$ is transitive.
6. Since K is transitive, $K \circ K \sqsubseteq K$. Hence $(K \circ K)(x,x) \leq K(x,x)$. That is, $\sqcup\{K(x,y) \sqcap K(y,x) \mid y \in S\} \leq K(x,x)$. Note that K is symmetric and hence $K(x,y) = K(y,x)$. Thus $\sqcup\{K(x,y) \sqcap K(x,y) \mid y \in S\} \leq K(x,x)$ and hence $K(x,y) \leq K(x,x)$. Once again, K being symmetric, $K(y,x) \leq K(x,x)$.
7. Let $(x,w), (w,z) \in K|_s^t$. Hence $K(x,w) \geq \langle s,t \rangle$ and $K(w,z) \geq \langle s,t \rangle$. Therefore, $K(x,w) = \sqcup\{K(x,y) \sqcap K(y,z) \mid y \in S\} \geq K(x,w) \sqcap K(w,z) \geq \langle s,t \rangle$. Thus $(x,w) \in K|_s^t$.

■

A c-relation K on a ciset H that is reflexive, symmetric, and transitive is called an equivalent c-relation on H.

We now proceed to investigate whether or not additional operations introduced in this chapter are precise.

FIGURE 6.1 Graph of F

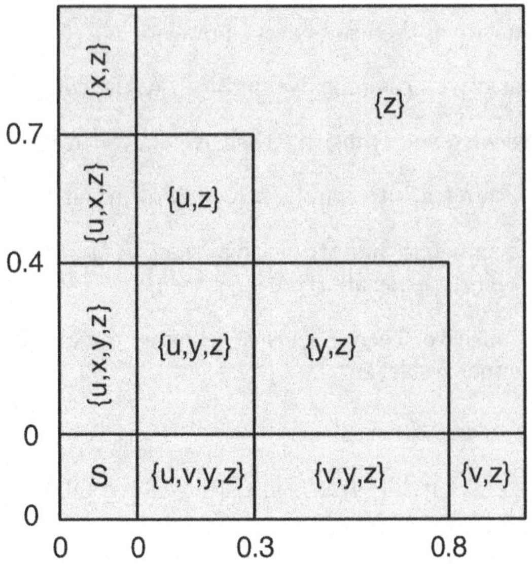

Example 6.3.12 *Let $S = \{u, v, x, y, z\}$ and let F be a ciset on S defined by*

$F(u) = \langle 0.7, 0.3\rangle, F(v) = \langle 0, 1\rangle, F(x) = \langle 1, 0\rangle, F(y) = \langle 0.4, 0.8\rangle, F(z) = \langle 1, 1\rangle$.

Then

$$F|_s^t = \begin{cases} S & s = 0; t = 0 \\ \{u, v, y, z\} & s = 0; 0 < t \leq 0.3 \\ \{v, y, z\} & s = 0; 0.3 < t \leq 0.8 \\ \{v, z\} & s = 0; 0.8 < t \leq 1 \\ \{u, x, y, z\} & 0 < s \leq 0.4; t = 0 \\ \{u, y, z\} & 0 < s \leq 0.4; 0 < t \leq 0.3 \\ \{y, z\} & 0 < s \leq 0.4; 0.3 < t \leq 0.8 \\ \{z\} & 0 < s \leq 0.4; 0.8 < t \leq 1 \\ \{u, x, z\} & 0.4 < s \leq 0.7; t = 0 \\ \{u, z\} & 0.4 < s \leq 0.7; 0 < t \leq 0.3 \\ \{z\} & 0.4 < s \leq 0.7; 0.3 < t \leq 1 \\ \{x, z\} & 0.7 < s \leq 1; t = 0 \\ \{z\} & 0.7 < s \leq 1; 0 < t \leq 1 \end{cases}$$

Thus a ciset F on S can be considered as a collection of subsets of S. The following fact is worth noticing.

F is completely determined by two descending chains of c-subsets of S, starting with S. In the above example, the descending chains are

$S \supset \{u, x, y, z\} \supset \{u, x, z\} \supset \{x, z\}$ and $S \supset \{u, v, y, z\} \supset \{v, y, z\} \supset \{v, z\}$. Also note that $F|_s^t = F|_s \cap F|^t$, for all $(s, t) \in [0, 1] \times [0, 1]$.

Proposition 6.3.13 *Let F be a ciset on a nonempty set S. Then for $0 \leq s, t \leq 1, F|_s^t = F|_s \cap F|^t$. Further, if the range of F is finite, there exists $0 = s_0 < \ldots < s_p = 1$ and $0 = t_0 < \ldots < t_q = 1$ such that $F|_{s_i} \supset F|_{s_{i+1}}, F|_{s_i} \neq F|_{s_{i+1}}$, for $i = 0, 1, \ldots, p-1$; and $F|^{t_j} \supset F|^{t_{j+1}}, F|^{t_j} \neq F|^{t_{j+1}}, j = 0, 1, \ldots, q-1$. Further, $F|_{s_0} = F|^{t_0} = S$.*

Example 6.3.14 *Consider the ciset F given in Figure 4.1. Observe that $s_0 = 0 < s_1 = 0.4 < s_2 = 0.7 < s_3 = 1$ and $F|_{s_0} = F|_0 = S \supset F|_{s_1} = F|_{0.4} = \{u, x, y, z\} \supset F|_{s_2} = F|_{0.7} = \{u, x, z\} \supset F|_{s_3} = F|_1 = \{x, z\} \supset \varnothing$. Thus $p = 3$. Similarly, $t_0 = 0 < t_1 = 0.3 < t_2 = 0.8 < t_3 = 1$ and $F|^{t_0} = F|^0 = S \supset F|^{t_1} = F|^{0.3} = \{u, v, y, z\} \supset F|^{t_2} = F|^{0.8} = \{v, y, z\} \supset F|^{t_3} = F|^1 = \{v, z\} \supset \varnothing$. Thus $q = 3$.*

From the above proposition, we see that a finite-valued ciset F on S determines two chains of c-subsets of S. Conversely, given two finite chains of c-subsets $S = C_0 \supset C_1 \supset \ldots \supset C_p, p > 0; S = D_0 \supset D_1 \supset \ldots \supset D_q, q > 0$; there exits a ciset F on S such that $C_i, i = 1, 2, \ldots, p$ are the set of lower cut sets of F and $D_j, j = 1, 2, \ldots, q$ are the set of upper cuts sets of F. The construction of F can be outlined as follows. Choose $p+q+2$ real numbers $s_i, i = 0, 1, \ldots, p$; $t_j, j = 0, 1, \ldots, q$ in $[0, 1]$ such that $0 = s_0 < \ldots < s_p = 1$ and $0 = t_0 < \ldots < t_q = 1$. For each $x \in S$,

$$F(x) = \begin{cases} \langle s_i, t_j \rangle & \text{if } x \in C_i - C_{i+1}, i = 0, 1, 2, \ldots, p-1 \\ & \text{and } x \in D_j - D_{j+1}, j = 0, 1, 2, \ldots, q-1; \\ \langle s_p, t_j \rangle & \text{if } x \in C_p \\ & \text{and } x \in D_j - D_{j+1}, j = 0, 1, 2, \ldots, q-1; \\ \langle s_i, t_q \rangle & \text{if } x \in C_i - C_{i+1}, i = 0, 1, 2, \ldots, p-1 \\ & \text{and } x \in D_q; \\ \langle s_p, t_q \rangle & \text{if } x \in C_p \\ & \text{and } x \in D_q. \end{cases}$$

Thus it follows that a finite-valued ciset is completely determined by two chains of c-subsets of S and set of values $0 = s_0 < \ldots < s_p = 1$ and $0 = t_0 < \ldots < t_q = 1$.

Example 6.3.15 *Let $p = q = 3$ and let $S = \{u, v, x, y, z\}$. Assume that $C_0 = S \supset C_1 = \{u, x, y, z\} \supset C_2 = \{u, x, z\} \supset C_3 = \{x, z\} \supset \varnothing$ and $D_0 = S \supset D_1 = \{u, v, y, z\} \supset D_2 = \{v, y, z\} \supset D_3 = \{v, z\} \supset \varnothing$. Further, let $s_0 = 0 < s_1 = 0.4 < s_2 = 0.7 < s_3 = 1$ and $t_0 = 0 < t_1 = 0.3 < t_2 = 0.8 < t_3 = 1$. Note that $u \in C_2 - C_3$ and $u \in D_1 - D_2$. Therefore $F(u) = \langle s_2, t_1 \rangle = \langle 0.7, 0.3 \rangle$. Similarly, $v \in C_0 - C_1$ and $v \in D_3$. Therefore $F(v) = \langle s_0, t_3 \rangle = \langle 0, 1 \rangle$. Now $x \in C_3$ and $x \in D_0 - D_1$. Therefore $F(u) = \langle s_3, t_0 \rangle = \langle 1, 0 \rangle$. Further $y \in C_1 - C_2$ and $y \in D_2 - D_3$. Therefore $F(y) =$*

FIGURE 6.2 Graph of G

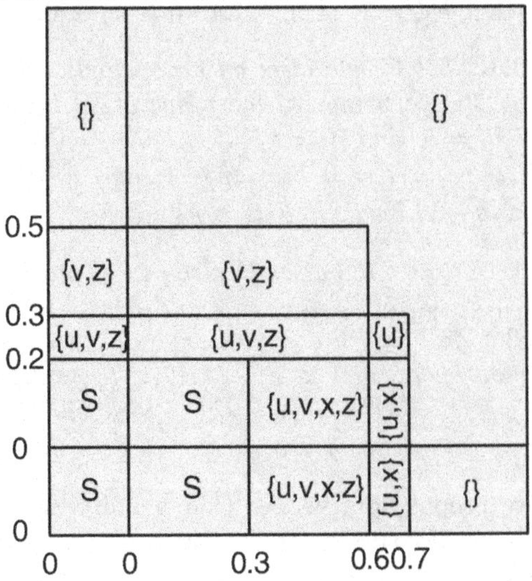

{} {}
0.5
{v,z} {v,z}
0.3
{u,v,z} {u,v,z} {u}
0.2
S S {u,v,x,z} {u,x}
0
S S {u,v,x,z} {u,x} {}
0
0 0 0.3 0.60.7

$\langle s_1, t_2\rangle = \langle 0.4, 0.8\rangle$. *Notice that* $z \in C_3$ *and* $z \in D_3$. *Therefore* $F(z) = \langle s_3, t_3\rangle = \langle 1, 1\rangle$.

Example 6.3.16 *Let* G *be a ciset on* S *defined by*

$G(u) = \langle 0.3, 0.7\rangle$, $G(v) = \langle 0.5, 0.6\rangle$, $G(x) = \langle 0.2, 0.7\rangle$, $G(y) = \langle 0.2, 0.3\rangle$, $G(z) = \langle 0.5, 0.6\rangle$.

Then $G_0 = S, G_{0.2} = S, G_{0.3} = \{u, v, z\}, G_{0.5} = \{v, z\}$ and $G_1 = \varnothing$. Similarly, $G^0 = S, G^{0.3} = S, G^{0.6} = \{u, v, x, z\}$ and $G^{0.7} = \{u, x\}$. Then

$$G|_s^t = \begin{cases} S & 0 \le s \le 0.2; 0 \le t \le 0.3 \\ \{u, v, x, z\} & 0 \le s \le 0.2; 0.3 < t \le 0.6 \\ \{u, x\} & 0 \le s \le 0.2; 0.6 < t \le 0.7 \\ \varnothing & 0 \le s \le 0.2; 0.7 < t \le 1 \\ \{u, v, z\} & 0.2 < s \le 0.3; 0 \le t \le 0.6 \\ \{u\} & 0.2 < s \le 0.3; 0.6 < t \le 0.7 \\ \varnothing & 0.2 < s \le 0.3; 0.7 < t \le 1 \\ \{v, z\} & 0.3 < s \le 0.5; 0 \le t \le 0.6 \\ \varnothing & 0.3 < s \le 0.5; 0.6 < t \le 1 \\ \varnothing & 0.5 < s \le 1; 0 \le t \le 1 \end{cases}$$

The ciset G on S can be considered as a collection of subsets of S.

Conversely, given two chains of subsets $C_0 = S \supset C_1 = \{u, v, z\} \supset C_2 = \{v, z\}, D_0 = S \supset D_1 = \{u, v, x, z\} \supset D_2 = \{u, x\}$ and real numbers $0.2 < 0.3 < 0.5, 0.3 < 0.6 < 0.7$, we notice that $p = 2$ and $q = 2$. We assign $s_0 = 0.2, s_1 = 0.3, s_2 = 0.5$. Similarly, $t_0 = 0.3, t_1 = 0.6, t_2 = 0.7$.

Note that $u \in C_1 - C_2$ and $u \in D_2$. Therefore $F(u) = \langle s_1, t_2 \rangle = \langle 0.3, 0.7 \rangle$. Similarly, $v \in C_2$ and $v \in D_1 - D_2$. Therefore $F(v) = \langle s_2, t_1 \rangle = \langle 0.5, 0.6 \rangle$. Now $x \in C_0 - C_1$ and $x \in D_2$. Therefore $F(u) = \langle s_0, t_2 \rangle = \langle 0.2, 0.7 \rangle$. Further $y \in C_0 - C_1$ and $y \in D_0 - D_1$. Therefore $F(y) = \langle s_0, t_0 \rangle = \langle 0.2, 0.3 \rangle$. Notice that $z \in C_2$ and $z \in D_1 - D_2$. Therefore $F(z) = \langle s_2, t_1 \rangle = \langle 0.5, 0.6 \rangle$.

In our above discussion, we have the following facts. Let H be a ciset. Then

- $t_0 = \wedge\{u(H(x)) \mid x \in S\}$,
- $t_q = \vee\{u(H(x)) \mid x \in S\}$,
- $s_0 = \wedge\{l(H(x)) \mid x \in S\}$,
- $s_p = \vee\{l(H(x)) \mid x \in S\}$.

The following theorem is quite clear and hence we state it without any proof.

Theorem 6.3.17 *Let F be a finite-valued ciset on a nonempty set S. Then there exists $0 = s_0 < \ldots < s_p = 1$; $0 = t_0 < \ldots < t_q = 1$ such that $F|_{s_p} \subset \ldots \subset F|_{s_0} = S$; $F|^{t_q} \subset \ldots \subset F|^{t_0} = S$ are lower and upper cut sets of F respectively. Conversely, given $p+q+2$ real numbers $s_i, i = 0, 1, \ldots, p$; $t_j, j = 0, 1, \ldots, q$ in $[0,1]$ such that $0 = s_0 < \ldots < s_p = 1$ and $0 = t_0 < \ldots < t_q = 1$.and two finite chains of c-subsets $S = C_0 \supset C_1 \supset \ldots \supset C_p, p > 0$; $S = D_0 \supset D_1 \supset \ldots \supset D_q, q > 0$; there exits a unique ciset F on S such that $C_i, i = 1, 2, \ldots, p$ are the set of lower cut sets of F and $D_j, j = 1, 2, \ldots, q$ are the set of upper cuts sets of F respectively.*

Given a ciset F, by the above Theorem, F is equivalent to an ordered pair $(\mathfrak{L}', \mathfrak{U}')$, where $\mathfrak{L}' = \{F|_s \mid s \in [0,1]\}$ is *chain of lower c-cut sets* of F and $\mathfrak{U}' = \{F|^t \mid t \in [0,1]\}$ is *chain of upper c-cut sets* of F respectively. Now define F^* as the set $\{F|_s \cap F|^t \mid F|_s, F|^t, (s,t) \in [0,1] \times [0,1]\}$.

Definition 6.3.18 *A ciset F is said to represent F^*. We indicate this as $rep(F) = F^*$. The set of subsets defined by $F^*(s,t) = F|_s \cap F|^t$, where $F_s \in \mathfrak{L}, F^t \in \mathfrak{U}, 0 \leq s \leq 1$ and $0 \leq t \leq 1$ is called the* alternate world *of F.*

If we define $F^*(s,t) = F|_s^t$ for $0 \leq s \leq 1$ and $0 \leq t \leq 1$, then $\{F^*(s,t) \mid 0 \leq s \leq 1, 0 \leq t \leq 1\}$ is the alternate world of F.

Example 6.3.19 *The ciset F presented in Example 6.3.14 in fact represents a collection of sets. They are $\{z\}$, $\{u,z\}$, $\{v,z\}$, $\{x,z\}$, $\{y,z\}$, $\{u,y,z\}$, $\{v,y,z\}$, $\{u,x,z\}$, $\{u,v,y,z\}$, $\{u,x,y,z\}$, S. Similarly, The ciset G presented in Example 6.3.16 represents the following collection of sets: $\varnothing$, $\{u\}$, $\{u,x\}$, $\{v,z\}$, $\{u,v,z\}$, $\{u,v,x,z\}$, S.*

FIGURE 6.3 Graph of F and G

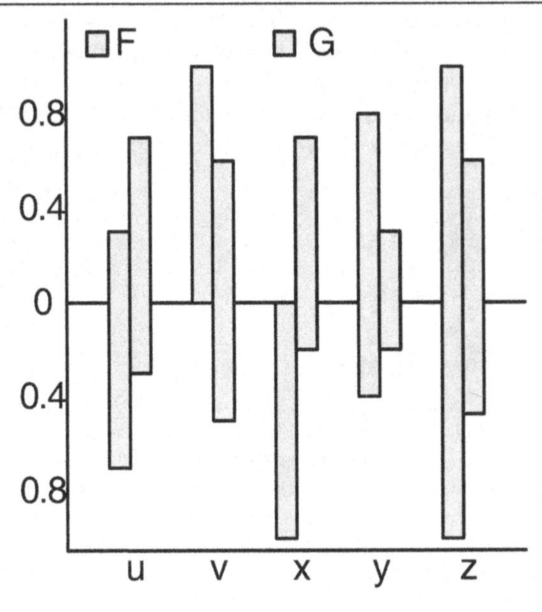

Having defined the notion of alternate world, we are now ready to investigate whether or not these operations are precise. It is also natural to graph a ciset, by treating it as a function. Thus the cisets F and G can be plotted as follows:

This presentation is quite useful in computing the ciset operations.

c-union

Example 6.3.20 *Consider the ciset F and G presented in Example 6.3.14 and in Example 6.3.16 respectively. It is quite easy to observe that*

$(F \sqcup G)(u) = \langle 0.7, 0.7 \rangle$,
$(F \sqcup G)(v) = \langle 0.5, 1.0 \rangle$,
$(F \sqcup G)(x) = \langle 1.0, 0.7 \rangle$,
$(F \sqcup G)(y) = \langle 0.4, 0.8 \rangle$
and
$(F \sqcup G)(z) = \langle 1.0, 1.0 \rangle$.

Recall that
$F|_{0.4} = \{u, x, y, z\}$,
$F|_{0.7} = \{u, x, z\}$,
$F|_{1} = \{x, z\}$.
Similarly,
$F|^{0.3} = \{u, v, y, z\}$,

FIGURE 6.4 Graph of $F \sqcup G$

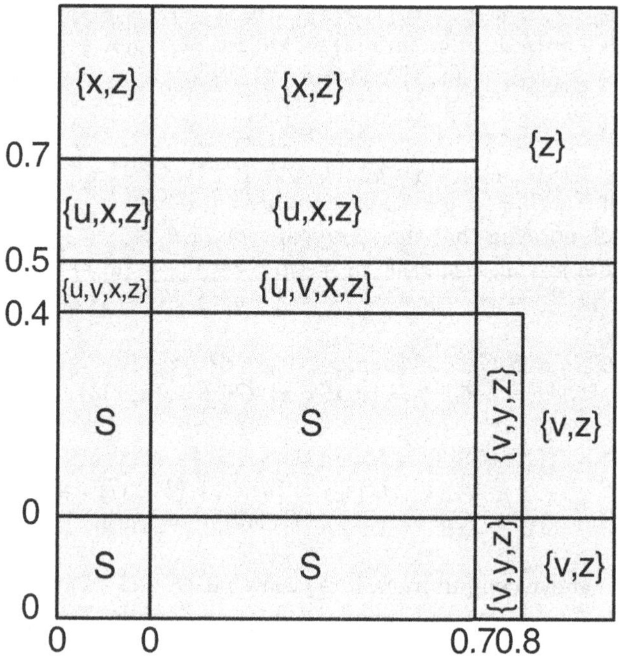

$F|^{0.8} = \{v, y, z\}$,
$F|^{1} = \{v, z\}$.
Also note that
$G|_{0.2} = S$,
$G|_{0.3} = \{u, v, z\}$,
$G|_{0.5} = \{v, z\}$.
Similarly,
$G|^{0.3} = S$,
$G|^{0.6} = \{u, v, x, z\}$,
$G|^{0.7} = \{u, x\}$.
Now
$(F \sqcup G)|_{0.4} = S = \{u, x, y, z\} \cup \{v, z\} = F|_{0.4} \cup G|_{0.4}$,
$(F \sqcup G)|_{0.5} = \{u, v, x, z\} = \{u, x, z\} \cup \{v, z\} = F|_{0.5} \cup G|_{0.5}$,
$(F \sqcup G)|_{0.7} = \{u, x, z\} = \{u, x, z\} \cup \varnothing = F|_{0.7} \cup G|_{0.7}$,
$(F \sqcup G)|_{1} = \{x, z\} = \{x, z\} \cup \varnothing = F|_{1} \cup G|_{1}$.
Further,
$(F \sqcup G)|^{1} = \{v, z\} = \{v, z\} \cup \varnothing = F|^{1} \cup G|^{1}$.
$(F \sqcup G)|^{0.8} = \{v, y, z\} = \{v, y, z\} \cup \varnothing = F|^{0.8} \cup G|^{0.8}$.
$(F \sqcup G)|^{0.7} = S = \{v, y, z\} \cup \{u, x\} = F|^{0.7} \cup G|^{0.7}$.
It is worth noticing that the representation of F is
$F^* = (\{F|_{0.4} = \{u, x, y, z\}, F|_{0.7} = \{u, x, z\}, F|_{1} = \{x, z\}\},$
$\{F|^{0.3} = \{u, v, y, z\}, F|^{0.8} = \{v, y, z\}, F|^{1} = \{v, z\}\})$.
Similarly,
$G^* = (\{G|_{0.2} = S, G|_{0.3} = \{u, v, z\}, G|_{0.5} = \{v, z\}\},$
$\{G|^{0.3} = S, G|^{0.6} = \{u, v, x, z\}, G|^{0.7} = \{u, x\}\})$.
Now
$(F \sqcup G)^* = (\{(F \sqcup G)|_{0.4} = S, (F \sqcup G)|_{0.5} = \{u, v, x, z\},$
$(F \sqcup G)|_{0.7} = \{u, x, z\}, (F \sqcup G)|_{1} = \{x, z\}\},$
$\{(F \sqcup G)|^{1} = \{v, z\}, (F \sqcup G)|^{0.8} = \{v, y, z\}, (F \sqcup G)|^{0.7} =$
$S\})$.

From the above computation, it is clear that $(F \sqcup G)^*$ can be computed from F^* and G^* using the (regular) union operation. Thus the c-union operation seems to be precise.

Theorem 6.3.21 *The binary operation c-union is precise.*

Proof. We shall prove that $(F \sqcup G)|_s = F|_s \cup G|_s, (F \sqcup G)|^t = F|^t \cup G|^t$ for $0 \leq s \leq 1, 0 \leq t \leq 1$. Let $x \in S$. Now $x \in (F \sqcup G)|_s$ if and only if $l((F \sqcup G)(x)) \geq s$ if and only if $l(F(x) \sqcup G(x)) \geq s$ if and only if $l(F(x)) \vee l(G(x)) \geq s$ if and only if $x \in F|_s$ or $x \in G|_s$ if and only if $x \in F|_s \cup G|_s$. Similarly, $x \in (F \cup G)|^t$ if and only if $x \in F|^t \cup G|^t$. ■

Note that above theorem do not prove that $(F \cup G)|_s^t = F|_s^t \cup G|_s^t$. In fact the result do not hold. Instead, we have the following:

Lemma 6.3.22 *Let F and G be two cisets. Then $(F \cup G)|_s^t \supset F|_s^t \cup G|_s^t$ for $0 \leq s \leq 1, 0 \leq t \leq 1$.*

Proof. $(F \sqcup G)|_s^t = (F \sqcup G)|_s \cap (F \sqcup G)|^t$
$= (F|_s \cup G|_s) \cap (F|^t \cup G|^t)$
$= (F|_s \cap F|^t) \cup (F|_s \cap G|^t) \cup (G|_s \cap F|^t) \cup (G|_s \cap G|^t)$
$= F|_s^t \cup (F|_s \cap G|^t) \cup (G|_s \cap F|^t) \cup G|_s^t$. ■

Example 6.3.23 *Consider the ciset F and G presented in Example 6.3.12 and in Example ?? respectively. Note that $F(y) = \langle 0.4, 0.8\rangle$ and $G(y) = \langle 0.2, 0.3\rangle$. Thus $(F \sqcup G)(y) = \langle 0.4, 0.8\rangle$. Therefore, we have the following:*

$y \in F|^{0.7}, y \notin F|_{0.5}, y \notin G|^{0.7}, y \in G|_{0.5}$.
Further $y \in (F \sqcup G)|_{0.5}^{0.7}$, $y \in (F \cup G)|_{0.5}$ and $y \in (F \cup G)|^{0.7}$.
Since, $y \notin F|_{0.5}^{0.7}$ and $y \notin G|_{0.5}^{0.7}$, $y \notin F|_{0.5}^{0.7} \cup G|_{0.5}^{0.7}$. Thus $y \in (F \cup G)|_{0.5}^{0.7} - (F|_{0.5}^{0.7} \cup G|_{0.5}^{0.7})$.

c-intersection

Example 6.3.24 *Consider the ciset F and G presented in Example 6.3.14 and in Example 6.3.16 respectively. It is quite easy to observe that*

$(F \sqcap G)(u) = \langle 0.3, 0.3\rangle$,
$(F \sqcap G)(v) = \langle 0, 0.6\rangle$,
$(F \sqcap G)(x) = \langle 0.2, 0\rangle$,
$(F \sqcap G)(y) = \langle 0.2, 0.3\rangle$
$(F \sqcap G)(z) = \langle 0.5, 0.6\rangle$.
Thus we have the following:
$(F \sqcap G)|_0 = S = S \cap S = F|_0 \cap G|_0$,
$(F \sqcap G)|_{0.2} = \{u, x, y, z\} = \{u, x, y, z\} \cap S = F|_{0.2} \cap G|_{0.2}$,
$(F \sqcap G)|_{0.3} = \{u, z\} = \{u, x, y, z\} \cap \{u, v, z\} = F|_{0.3} \cap G|_{0.3}$,
$(F \sqcap G)|_{0.5} = \{z\} = \{u, x, z\} \cap \{v, z\} = F|_{0.5} \cap G|_{0.5}$.
Also note that
$(F \sqcap G)|^0 = S = S \cap S = F|^0 \cap G|^0$,
$(F \sqcap G)|^{0.3} = \{u, v, y, z\} = \{u, v, y, z\} \cap S = F|^{0.3} \cap G|^{0.3}$,
$(F \sqcap G)|^{0.6} = \{v, z\} = \{v, y, z\} \cap \{u, v, x, z\} = F|^{0.6} \cap G|^{0.6}$.
It is worth noticing that the representation of F is
$F^* = (\{F|_{0.4} = \{u, x, y, z\}, F|_{0.7} = \{u, x, z\}, F|_1 = \{x, z\}\},$
$\{F|^{0.3} = \{u, v, y, z\}, F|^{0.8} = \{v, y, z\}, F|^1 = \{v, z\}\})$.
Similarly,
$G^* = (\{G|_{0.2} = S, G|_{0.3} = \{u, v, z\}, G|_{0.5} = \{v, z\}\},$
$\{G|^{0.3} = S, G|^{0.6} = \{u, v, x, z\}, G|^{0.7} = \{u, x\}\})$.
Now
$(F \sqcap G)^* = (\{(F \sqcap G)|_0 = S, (F \sqcap G)|_{0.2} = \{u, x, y, z\},$
$(F \sqcap G)|_{0.3} = \{u, z\}, (F \sqcap G)|_{0.5} = \{z\}\},$
$\{(F \sqcap G)|^0 = S, (F \sqcap G)|^{0.3} = \{u, v, y, z\}, (F \sqcap G)|^{0.6} = \{v, z\}\})$.

FIGURE 6.5 Graph of $F \sqcap G$

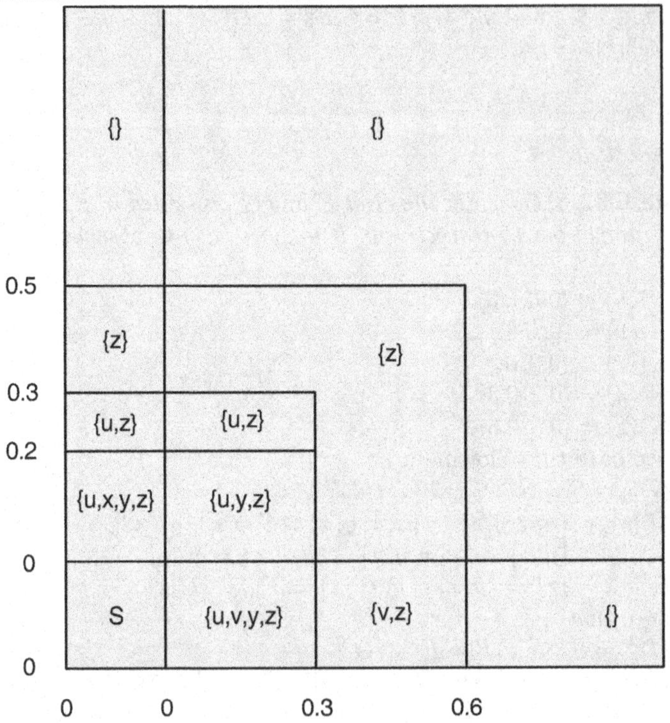

{}
{}
0.5
{z}
{z}
0.3
{u,z}
{u,z}
0.2
{u,x,y,z}
{u,y,z}
0
S
{u,v,y,z}
{v,z}
{}
0
0
0
0.3
0.6

From the above computation, it is clear that $(F \sqcap G)^*$ can be computed from F^* and G^* using the (regular) union operation. Thus the c-intersection operation seems to be precise.

Theorem 6.3.25 *The binary operation c-intersection is precise.*

Proof. We shall prove that $(F \sqcap G)|_s = F|_s \cap G|_s$, $(F \sqcap G)|^t = F|^t \cap G|^t$ for $0 \leq s \leq 1, 0 \leq t \leq 1$. Let $x \in S$. Now $x \in (F \sqcap G)|^t$ if and only if $u((F \sqcap G)(x)) \geq t$ if and only if $u(F(x) \sqcap G(x)) \geq t$ if and only if $u(F(x)) \wedge u(G(x)) \geq t$ if and only if $x \in F|^t$ and $x \in G|^t$ if and only if $x \in F|^t \cap G|^t$. Similarly, $(F \sqcap G)|_s = F|_s \cap G|_s$. ■

Note that above theorem in fact prove that $(F \sqcap G)|_s^t = F|_s^t \cap G|_s^t$.

Corollary 6.3.26 *Let F and G be two cisets. Then $(F \sqcap G)|_s^t = F|_s^t \cap G|_s^t$ for $0 \leq s \leq 1, 0 \leq t \leq 1$.*

Proof. $(F \sqcap G)|_s^t = (F \sqcap G)|_s \cap (F \sqcap G)|^t$
$= (F|_s \cap G|_s) \cap (F|^t \cap G|^t)$
$= (F|_s \cap F|^t) \cap (G|_s \cap G|^t)$
$= F|_s^t \cap G|_s^t$. ■

c-negation

Next we proceed to show that the binary ciset c-negation operation is not precise.

Example 6.3.27 *Consider the ciset G presented in Example 6.3.16 respectively. Recall that,*
$G(u) = \langle 0.3, 0.7 \rangle$, $G(v) = \langle 0.5, 0.6 \rangle$, $G(x) = \langle 0.2, 0.7 \rangle$,
$G(y) = \langle 0.2, 0.3 \rangle$, $G(z) = \langle 0.5, 0.6 \rangle$ *and*
$G^* = (\{G|_{0.2} = S, G|_{0.3} = \{u, v, z\}, G|_{0.5} = \{v, z\}\},$
$\{G|^{0.3} = S, G|^{0.6} = \{u, v, x, z\}, G|^{0.7} = \{u, x\}\})$.
Note that $\neg G(u) = \langle 0.7, 0.3 \rangle$, $\neg G(v) = \langle 0.5, 0.4 \rangle$,
$\neg G(x) = \langle 0.8, 0.3 \rangle$, $\neg G(y) = \langle 0.8, 0.7 \rangle$, $\neg G(z) = \langle 0.5, 0.4 \rangle$.
Further, $\neg G|_{0.8} = \{x, y\}$, $\neg G|_{0.7} = \{u, x, y\}$ *and* $\neg G|_{0.5} = S$.
Now
$\neg G|_{0.7} = \{u, x, y\} = S - \{v, z\} = \{v, z\}^c \neq (G|_{0.7})^c$ *or* $(G|^{0.7})^c$.
Thus unary operation $\neg$ *on ciset is not precise.*

c-difference

Example 6.3.28 *Consider the ciset F and G presented in Example 6.3.14 and in Example 6.3.16 respectively. Recall that,*
$(F \neg G)(u) = \langle 0.7, 0.3 \rangle \neg \langle 0.3, 0.7 \rangle = \langle 0.7, 0 \rangle$,
$(F \neg G)(v) = \langle 0, 1 \rangle \neg \langle 0.5, 0.6 \rangle = \langle 0, 1 \rangle$,
$(F \neg G)(x) = \langle 1, 0 \rangle \neg \langle 0.2, 0.7 \rangle = \langle 1, 0 \rangle$,

$(F\neg G)(y) = \langle 0.4, 0.8\rangle \neg \langle 0.2, 0.3\rangle = \langle 0.4, 0.8\rangle$,
$(F\neg G)(z) = \langle 1, 1\rangle \neg \langle 0.5, 0.6\rangle = \langle 1, 1\rangle$.
Now,
$(F\neg G)|_0 = S, (F\neg G)|_{0.4} = \{u, x, y, z\}, (F\neg G)|_{0.7} = \{u, x, z\}, (F\neg G)|_1 = \{x, z\}$. *Also,* $(F\neg G)|^0 = S, (F\neg G)|^{0.8} = \{v, y, z\}, (F\neg G)|^1 = \{v, z\}$. *Now,* $F|_0 = S, F|_{0.4} = \{u, x, y, z\}, F|_{0.7} = \{u, x, z\}$ *and* $F|_1 = \{x, z\}$. *Similarly,* $F|^1 = S, F|^{0.3} = \{u, v, y, z\}, F|^{0.8} = \{v, y, z\}$ *and* $F|^1 = \{v, z\}$. *Recall that* $G|_0 = S, G|_{0.2} = S, G|_{0.3} = \{u, v, z\}, G|_{0.5} = \{v, z\}$ *and* $G|_1 = \varnothing$. *Similarly,* $G|^0 = S, G|^{0.3} = S, G|^{0.6} = \{u, v, x, z\}, G|^{0.7} = \{u, x\}$ *and* $G|^1 = \varnothing$ *It is easy to see that* $(F\neg G)|^{0.4} = \{v, y, z\}$ *and* $F|^{0.4}\neg G|^{0.4} = \{v, y, z\}\neg\{u, v, x, z\} = \{y\}$. *Thus* $(F\neg G)|^{0.4} \neq F|^{0.4}\neg G|^{0.4}$.

Thus the operation c-difference is not precise.

c-Cartesian product

Theorem 6.3.29 *The binary ciset theory operation c-Cartesian product is precise.*

The proof is similar to that the proof that Cartesian product is precise and hence omitted.

7

INFORMATION SOURCE TRACKING

The problem of representing and processing inaccurate data is the subject of the study in [31, 32, 33, 34, 35]. The approach used was not only to keep tack of information, but also to keep track of sources confirming a piece of data along with the data and use that additional information in all data manipulation activities. In particular, the information about the source is used to query data. To this end, author had extended the relational algebra operators and used those extended operators to produce the answer to the query as well as the information regarding the sources contributed to the answers to the query.

In this chapter, we presented the main ideas contained in this approach. There are two main reasons for us to have a closer look at this approach. First, it is an important work in the sense that no one else has carried out a similar work that can keep track of the sources contributing the data. Thus, the approach is major development in the area of multi-source database. Secondly, we can incorporate the ideas presented to further extend the ciset relational database. In the case of a ciset relational database, up to this point, we have not advocated source tracking, rather, we proposed collecting information from various sources and integrating them. Thus at this point we are treating all sources the same way and as such no effort is being made to keep track of the sources. Thus the approach taken in [31, 32, 33, 34, 35] and the approach presented in this book so far are orthogonal and as such both the approaches can be applied simultaneously.

7.1 The IST Model

Just as in ciset relational model, from the user perspective, database model is still the classical relational model. The only conceptual difference as far as the user is concerned can be summarized as follows:

1. As user insert a new piece of data, user is asked to identify the contributing information sources as well. Similar is the situation when the user try to delete a piece of information or modify a piece of information.

2. In addition to producing results of a query, system identifies the contributing information sources for each of the tuples in the result of the query. The database can further calculate, for each tuple in the answer to the query, the probability of the validity the tuple. This probability is a function of the probability of the correctness of contributing information sources, which should be supplied by users of the system. Above all, user will be given the option to process the query as in a relational database and thus completely ignore the information source tracking feature of the system during query processing. However, user must provide information about contribution sources during the insertion, deletion or modification of data.

Now we proceed to describe the system's point of view. The database model is in fact an extended relational model as explained below. An *extended relation* scheme R is a set of attributes $\{A_1, A_2, \ldots, A_n, I\}$ where $A_1, A_2, \ldots, A_n$ relational attributes, and I is a special attribute called information source attribute or source attribute for short. Each attribute A_i has a domain of values $D_i, i = 1, 2, \ldots, n$. The domain of the source attribute I, denoted by D_I, is the set of vectors of length k with $-1, 0, 1$ elements, that is,

$$D_I = \{(a_1, a_2, \ldots, a_k) \mid a_i \in \{-1, 0, 1\}, i = 1, 2, \ldots, k\}$$

where k is the number of information sources we are tracking in this database. An element of the set D_I is called an *information source vector* or *source vector* for short. When there is no confusion, a source vector such as $(a_1, a_2, \ldots, a_k)$ will be represented as $(a_1 a_2 \ldots a_k)$. Further, we shall use the notation $\bar{1}$ in place of -1.

A tuple on the extended relational scheme $R = \{A_1, A_2, \ldots, A_n, I\}$ is an element of $D_1 \times D_2 \times \ldots \times D_n \times D_I$. Following the Sadri's lead, we also use the notation $t@u$ to denote a tuple on the extended scheme R, where t is a tuple on the relational scheme $\{A_1, A_2, \ldots, A_n\}$ and u is a k tuple which is an element of the set D_I, corresponding to the source attribute I of the extended relation scheme R. Note that k, the length of the source vector, is the number of different sources we are interested in tracking. We call t a *pure tuple* and u the source vector corresponding to t. A *extended relation instance* (or *extended relation* for short) r on the extended relation scheme R is a set of tuples on R.

Example 7.1.1 *Consider the following extended relation:*

TABLE 7.1 extended relation FACULTY

FACULTY

F_ID	FNAME	DEPT	I
72356	James Alt	Marketing	010
90897	Rose Lewis	Mathematics	001
91867	Mary Hill	Computer Science	010
43720	Jack Fox	Accounting	100

The extended relational scheme is $\{F\_ID, FNAME, DEPT, I\}$.
$DOM(F\_ID)$ *is the set of positive integers*
$DOM(FNAME)$ *is the set of names*
$DOM(DEPT)$ *is the set of department names*
$DOM(I) = \{\bar{1}, 0, 1\} \times \{\bar{1}, 0, 1\} \times \{\bar{1}, 0, 1\}$.

We use the notation $t@u$ to represent extended tuples such as (72356, James Alt, Marketing, 010). In this case, the pure tuple t is the relational tuple (72356, James Alt, Marketing) and u is the source vector $(0, 1, 0)$.

Semantic interpretation of source vectors

Let $R = \{A_1, A_2, \ldots, A_n, I\}$ be an extended relational scheme and let r be an instance of R. Now, consider a pure tuple t in r. The source vector corresponding to t identify information sources that are contributing either positively or negatively to the pure tuple t. In this model as presented in [35], as opposed to ciset relational model presented in Chapter 3, there can be more than one occurrence of the pure tuple in r. In other words, a pure tuple can have many source vectors associated with it. Thus in general, a pure tuple t may have p ($p > 0$) source vectors associated with it, say $u_1, u_2, \ldots, u_p$. In other words, there can be $p > 1$ tuples $t@u_1$, $t@u_2, \ldots, t@u_p$ in r. We now proceed to introduce the concept of an *expression* corresponding to a pure tuple t of r. This concept is central to the semantic interpretation of source vectors.

Let us first consider the case where $p = 1$. In this case there exist a unique extended tuple associated with a pure tuple t. Let us denote the extended tuple by $t@u$, where $u = (a_1, a_2, \ldots, a_k)$. We also denote the set of information sources by $S = \{s_1, s_2, \ldots, s_k\}$, where k is the number of information sources. Let us denote the set of information sources contributing positively to the pure tuple t by $S^+ = \{s_i \in S \mid a_i = 1\}$. Similarly, the set of information sources contributing negatively to the pure tuple t by $S^- = \{s_i \in S \mid a_i = \bar{1}\}$. Further, for each source $s_i \in S$, associate a Boolean variable f_i. The expression corresponding to t with respect u, denoted by $e(t@u)$, is written as

$$e(t@u) = \bigwedge_{s_i \in S+} f_i \bigwedge_{s_i \in S-} \neg f_i$$

Next we consider the case where $p > 1$. That is, $t@u_1, t@u_2, \ldots, t@u_p$ are all tuples with the the same pure part t in r. In this case, we use an alternate notation $t@z \in r$ where $z = \{u_1, u_2, \ldots, u_p\}$. The expression corresponding to t with respect to z is written as

$$e(t) = e(t@z) = \bigvee_{i=1}^{p} e(t@u_i)$$

The expression corresponding to a tuple t in a relation r as a propositional logic expression, where $f_1, f_2, \ldots, f_k$ represent Boolean variables. A truth assignment $f_i = true$ is interpreted as "information source i is believed to be correct or reliable". On the other hand, a truth assignment $f_i = false$ is interpreted as "information source i is believed to be incorrect or unreliable". Thus the truth value of $e(t)$ is a function of the truth values of $f_1, f_2, \ldots, f_k$, and indicates whether or not the tuple t is a valid or reliable tuple depending upon $e(t) = true$ or $false$ respectively.

The expression corresponding to a tuple t in a relation r can also be used to derive probabilistic information about t, that is, given probabilities for correctness of sources, $s_1, s_2, \ldots, s_k$ we can calculate the probability of the validity of t. This issue will be discussed in detail, later in this chapter. We could also use confidence indexes in place of just true or false values or probabilities. We shall address this variation in more detail later on.

Example 7.1.2 *Consider the following extended relation:*

TABLE 7.2 extended relation FACULTY

FACULTY

F_ID	FNAME	DEPT	I
12312	John Smith	Marketing	100
31897	Mary Lee	Mathematics	100
31897	Mary Lee	Mathematics	010
56739	Bea Anthony	Accounting	001

The expression corresponding to pure tuple (12312, John Smith, Marketing) is f_1. Similarly, the expression corresponding to the pure tuple (31897, Mary Lee, Mathematics) is given by $f_1 \vee f_2$. Further, expression corresponding to pure tuple (56739, Bea Anthony,Accounting) is f_3.

Note that the above extended relation is semantically not equivalent to extended relation

TABLE 7.3 FCLTY extended relation

FCLTY

FID	FNAME	DEPT	I
12312	John Smith	Marketing	1 0 0
31897	Mary Lee	Mathematics	1 1 0
56739	Bea Anthony	Accounting	0 0 1

extended relation in the sense that expressions for each of the pure tuples is the same in both relations. In FACULTY the expression corresponding to the pure tuple (31897, Mary Lee, Mathematics) is given by $f_1 \vee f_2$. Whereas, in FCLTY the expression corresponding to the pure tuple (31897, Mary Lee, Mathematics) is given by $f_1 \wedge f_2$. Thus FACULTY is semantically not equivalent to FCLTY. This further clarifies the need for multiple tuples in an extend relation corresponding to one pure tuple in this model.

It is worth noticing that $\bar{1}$ do not appear in source vectors of base relations. They can appear in computed relations such as views and answers to queries. To be more specific, the set difference operation in relational algebra and the not exists construct of SQL both produce $\bar{1}$ entries in computed relations. Note that IST model just tracks the information sources and do not try to integrate the positive and negative facts dealing with the same entity as explained in Chapter 3.

Example 7.1.3 *Consider the following extended relations of students and employees of a university:*

TABLE 7.4 extended relations EMPLOYEE and STUDENT

EMPLOYEE

ID	ENAME	DEPT	I
14265	Mark Daub	Marketing	1 0 0
18945	Lee Wong	Mathematics	1 0 0
18945	Lee Wong	Mathematics	0 1 0
73912	Cathy Cox	Accounting	0 0 1

STUDENT

ID	ENAME	MAJOR	I
14265	Mark Daub	Marketing	0 1 0
23426	Kris Lamb	English	1 0 0
84909	Cory Work	Music	0 1 0
73912	Cathy Cox	Accounting	0 1 0

The answer to the query "Find all students who are not employees" will yield the following computed extended relation. (We show how the answer is arrived in due course. For the present, assume that answer can be computed and is as given below.)

ID	*ENAME*	*MAJOR*	*I*
14265	*Mark Daub*	*Marketing*	$\bar{1}\,1\,0$
23426	*Kris Lamb*	*English*	1 0 0
84909	*Cory Work*	*Music*	0 1 0
73912	*Cathy Cox*	*Accounting*	$0\,1\,\bar{1}$

The expression corresponding to pure tuple (14265, Mark Daub, Marketing) is $\neg f_1 \wedge f_2$. *Therefore, Mark is in the list if and only if source one is incorrect and source two is correct. Similarly, the expression corresponding to pure tuple (73912, Cathy Cox, Accounting) is* $\neg f_3 \wedge f_2$. *As a consequence, Cathy is in the list if and only if source two is correct and source three is incorrect.*

7.2 Alternate Worlds

In this section we formalize the semantics of the extended relational model presented in this chapter. As indicated in Chapter 4, the closely related notions of representation, possibility functions, alternate worlds have been the tools used by leading researchers to formalize the information content of a databases with incomplete information [1, 3, 7, 9, 11, 12, 14, 17, 22, 23, 24, 28, 35, 43, 44]. The approach we adopt in this instance is similar to theirs. To be specific, we use the notion of alternate worlds to formalize the information content of a extended relational database. An extended relation represents a set of (regular) relations. Once this set has been completely identified, a query on extended relations can be as well processed against the set of (regular) relations represented by extended relations involved. Once again, we note that this approach is computationally inefficient and is not an approach we would recommend. On the other hand, this approach will very well explain the semantics in a formal setting. From a practical point of view, what we need is a query processing methodology that can be applied directly to extended relations. Further, we would like to generate the same answers that would be obtained by processing the query against represented (regular) relations. A query processing methodology that satisfies the above condition is called precise.

In the following presentation, we will characterize the set of (regular) relations represented by an extended relation.

Given an extended relation r on the extended scheme R,
$r = \{t_1@z_1, t_2@z_2, \ldots, t_n@z_n\}$
where $z_1, z_2, \ldots, z_n$ are sets of source vectors, we define r^* as a function from the set of subsets of information sources $S = \{s_1, s_2, \ldots, s_k\}$ to the set of (regular) relations *Rel* on the scheme $R - \{I\}$, that is,

$$r^* : 2^S \rightarrow Rel$$

Let Q be a subset of S, the set of information sources. Assign truth value *true* to members of Q and assign truth value *false* to members of S that are not in Q. Let us denote this truth assignment by $truth(Q)$. Then

$$r^*(Q) = \{t_i \mid e(t_i) = true \text{ under } truth(Q)\}.$$

In the above equation, $e(t_i)$ is the expression of t_i in r.

Definition 7.2.1 *An extended relation r represents the function r^*. We write this as $rep(r) = r^*$. The set of (regular) relations $r^*(Q)$, $Q \subseteq S$, is called the alternate world of r.*

Informally, an extended relation r represents the set of (regular) relations consisting of those tuples that would be valid if the sources in Q were correct or reliable and all other sources were incorrect or not reliable for all $Q \subseteq S$.

Example 7.2.2 *Consider the extended relation FACULTY first presented in Example 7.1.2. We present the table here once more for convenience sake.*

FACULTY

F_ID	*FNAME*	*DEPT*	*I*
12312	*John Smith*	*Marketing*	1 0 0
31897	*Mary Lee*	*Mathematics*	1 0 0
31897	*Mary Lee*	*Mathematics*	0 1 0
56739	*Bea Anthony*	*Accounting*	0 0 1

This relation in fact represents eight (regular) relations corresponding to eight different subsets of S. Note that there are three sources; say s_1, s_2, s_3. Thus there are eight subsets:

$\emptyset, \{s_1\}, \{s_2\}, \{s_3\}, \{s_1, s_2\}, \{s_2, s_3\}, \{s_1, s_3\}, S.$

The relation corresponding to the empty set is an empty relation. The relations corresponding to the other seven relations are shown below in order.

F_ID	*FNAME*	*DEPT*
12312	*John Smith*	*Marketing*
31897	*Mary Lee*	*Mathematics*

F_ID	*FNAME*	*DEPT*
31897	*Mary Lee*	*Mathematics*

F_ID	*FNAME*	*DEPT*
56739	*Bea Anthony*	*Accounting*

F_ID	FNAME	DEPT
12312	John Smith	Marketing
31897	Mary Lee	Mathematics

F_ID	FNAME	DEPT
31897	Mary Lee	Mathematics
56739	Bea Anthony	Accounting

F_ID	FNAME	DEPT
12312	John Smith	Marketing
31897	Mary Lee	Mathematics
56739	Bea Anthony	Accounting

F_ID	FNAME	DEPT
12312	John Smith	Marketing
31897	Mary Lee	Mathematics
56739	Bea Anthony	Accounting

Recall the concept of precision of extended operation. For the sake of convenience, we reproduce those definitions.

Definition 7.2.3 *Let $\otimes$ be a unary relational operation. A unary extended relational operation $\otimes'$ is said to be precise, if*

$$rep(\otimes'(r)) = \otimes(rep(r))$$

for all extended relations r, where $\otimes(rep(r))$ represents a function f such that $f(Q) = \otimes(r^*(Q))$ for all $Q \subseteq S$.

The above definition can be graphically depicted as follows:

$$\begin{array}{ccc} r & \xrightarrow{\otimes'} & \otimes'(r) \\ rep \downarrow & & \downarrow rep \\ rep(r) & \xrightarrow{\otimes} & rep(\otimes'(r)) = \otimes(rep(r)) \end{array}$$

Definition 7.2.4 *Let $\otimes$ be a binary relational operation. A binary extended relational operation $\otimes'$ is said to be precise, if*

$$rep(r \otimes' s) = rep(r) \otimes rep(s)$$

for all extended relations r, s where $rep(r) \otimes rep(s)$ represents a function f such that $f(Q) = r^*(Q) \otimes s^*(Q)$ for all $Q \subseteq S$.

The above definition can be graphically depicted as follows:

$$\begin{array}{ccc} (r,s) & \xrightarrow{\otimes'} & r \otimes' s \\ rep \big\downarrow & & \big\downarrow rep \\ (rep(r), rep(s)) & \xrightarrow{\otimes} & rep(r \otimes' s) = rep(r) \otimes rep(s) \end{array}$$

We now proceed to present the extended relational algebra operations and prove that they are precise.

7.3 Extended Relational Algebra Operations

In this section we extend relational algebra operations union, selection, projection, Cartesian product, natural join, and set difference as defined in [35] to manipulate source vectors. These extensions provide the necessary tools to identify source(s) contributing to each tuple in the answer to a query, and to calculate the reliability of a tuple in the answer as a function of the reliabilities of information sources.

We begin this section with operations on the source vectors. This will allow us to present extended operations in a formal setting. We conclude this section proving that the extended relational algebra operations of [35] are precise.

Operations on source vectors

We introduce three operations on source vectors. They are 3OR, NEGATION, and UNION [35]. The 3OR of two non-zero source vectors, $u = (a_1 ... a_k)$ and $v = (b_1 ... b_k)$ is a source vector $w = (c_1 ... c_k)$ obtained using the following look up table:

a_i	b_i	c_i
$\bar{1}$	$\bar{1}$	$\bar{1}$
$\bar{1}$	0	$\bar{1}$
$\bar{1}$	1	$w = 0$
0	$\bar{1}$	$\bar{1}$
0	0	0
0	1	1
1	$\bar{1}$	$w = 0$
1	0	1
1	1	1

where $w = 0$ indicates that the whole w vector is zero. We write $w = u \| v$ to indicate the 3OR operation. Let $u = (01\bar{1}0)$ and $v = (\bar{1}01\bar{1})$. then $u \| v$ $= (\bar{1}1\bar{1}1)$. Note that if we restrict the values of a_i and b_i in the table to Boolean values, 0 and 1, then we obtain the following:

a_i	b_i	c_i
0	0	0
0	1	1
1	0	1
1	1	1

This is the Boolean OR operation where 0 represents false and 1 represents true. Similarly, if we restrict the values of a_i and b_i in the table to Boolean values, 0 and $\bar{1}$, then we obtain the following:

a_i	b_i	c_i
$\bar{1}$	$\bar{1}$	$\bar{1}$
$\bar{1}$	0	$\bar{1}$
0	$\bar{1}$	$\bar{1}$
0	0	0

This is once again the Boolean OR operation where 0 represents false and $\bar{1}$ represents true. Thus 3OR is in fact an extension of our traditional Boolean OR; and hence the name 3OR.

Thus both 1 and $\bar{1}$ behave identically as far as 3OR is concerned. This is an important feature of 3OR.

Lemma 7.3.1 *The binary operation 3OR is commutative, associative and* (0...0) *is the identity and there exists* $2^r - 1$ *inverses where* r *denote the number of non-zero components of a source vector.* ■

Let $v_0 = (\bar{1}0\bar{1}1)$. then $v_1 = (10\bar{1}1), v_2 = (\bar{1}011), v_3 = (\bar{1}0\bar{1}\bar{1}), v_4 = (1011), v_5 = (10\bar{1}\bar{1}), v_6 = (\bar{1}01\bar{1}), v_7 = (101\bar{1})$.are the seven inverses of v. The following fact is worth noticing:

	v_0	v_1	v_2	v_3	v_4	v_5	v_6	v_7
v_0	v	w	w	w	w	w	w	w
v_1	w	v_1	w	w	w	w	w	w
v_2	w	w	v_2	w	w	w	w	w
v_3	w	w	w	v_3	w	w	w	w
v_4	w	w	w	w	v_4	w	w	w
v_5	w	w	w	w	w	v_5	w	w
v_6	w	w	w	w	w	w	v_6	w
v_7	w	w	w	w	w	w	w	v_7

In other words, $v_0, v_1, \ldots, v_7$ are mutually inverses of each other.

The 3OR of two sets of source vectors, $x = \{u_1, ..., u_p\}$ and $y = \{v_1, ..., v_q\}$, is calculated pairwise. That is,

$$x \parallel y = \{u_1 \parallel v_1, ..., u_1 \parallel v_q, ..., u_p \parallel v_1, ..., u_p \parallel v_q\}$$

It is worth noticing that if x and y has p and q elements respectively, then $x \parallel y$ may have at most $p \times q$ elements. In many situations it may not produce $p \times q$ elements. This can be illustrated through the following example.

Example 7.3.2 *Let* y, z *be two source vectors as given below.*

$$y = \{(1000), (0010), (000\bar{1})\}$$

and

$$z = \{(1000), (00\bar{1}0), (000\bar{1})\}$$

Then $y \parallel z$

$= \{(1000) \parallel (1000), (0010) \parallel (1000), (000\bar{1}) \parallel (1000), (1000) \parallel (00\bar{1}0),$
$(0010) \parallel (00\bar{1}0), (000\bar{1}) \parallel (00\bar{1}0), (1000) \parallel (000\bar{1}), (0010) \parallel (000\bar{1}),$
$(000\bar{1}) \parallel (000\bar{1})\}$
$= \{(1000), (1010), (100\bar{1}), (10\bar{1}0), (0000), (00\bar{1}\bar{1}), (100\bar{1}), (001\bar{1}), (000\bar{1})\}$
$= \{(1000), (1010), (100\bar{1}), (10\bar{1}0), (0000), (00\bar{1}\bar{1}), (001\bar{1}), (000\bar{1})\}$

Note that in this case, $y \parallel z$ *has only eight elements; not nine elements.*

Theorem 7.3.3 *Given* $t_1 = t@x, t_2 = t@y$, *and* $t_3 = t@z$, *where* x, y, *and* z *are sets of source vectors, if* $x = y \parallel z$ *then* $e(t_1) = e(t_2) \wedge e(t_3)$, *where* $e(t_i)$ *denotes the expression corresponding to* t_i. ■

Example 7.3.4 *Let* $y = \{(1000), (0010), (000\bar{1})\}$ *and* $z = \{(1000), (00\bar{1}0), (000\bar{1})\}$ *be two source vectors. Then*

$e(t_1) = (f_1) \vee (f_1 \wedge f_3) \vee (f_1 \wedge \neg f_4) \vee (f_1 \wedge \neg f_3) \vee (\neg f_3 \wedge \neg f_4) \vee (f_3 \wedge \neg f_4) \vee (\neg f_4)$.

Now,

$e(t_2) = (f_1) \vee (f_3) \vee (\neg f_4)$

and

$e(t_3) = (f_1) \vee (\neg f_3) \vee (\neg f_4)$

Therefore,

$e(t_2) \wedge e(t_3) = ((f_1) \vee (f_3) \vee (\neg f_4)) \wedge ((f_1) \vee (\neg f_3) \vee (\neg f_4))$
$= (f_1) \vee (f_1 \wedge f_3) \vee (f_1 \wedge \neg f_4) \vee (f_1 \wedge \neg f_3) \vee (\neg f_3 \wedge \neg f_4) \vee (f_3 \wedge \neg f_4) \vee (\neg f_4)$
$= e(t_1)$.

The NEGATION is a unary operation on source vectors and is defined as follows: Let $u = (a_1 ... a_k)$ be a source vector, and let $a_i, ..., a_{i_n}$ be the non-zero elements of u. The NEGATION of u, written $\#(u)$, is a set of source vectors $\{v_{i_1}, ..., v_{v_n}\}$ constructed as follows: All the elements of v_{i_j} are zero except its i_jth element, which is $1(\bar{1})$ if $a_{i_j} = \bar{1}(1)$.

Example 7.3.5 *Let* $u = (1\bar{1}01)$, *then* $\#(u) = \{(\bar{1}000), (0100), (000\bar{1})\}$.

The NEGATION of a set of source vectors $x = \{u_1, ..., u_p\}$ *is calculated as follows:*

$\#(x) = \#(u_1) \parallel ... \parallel \#(u_p)$

Example 7.3.6 *Let* $y = \{(1000), (0010), (000\bar{1})\}$ *be a source vector. Then*

$\#(y) = \#(1000) \parallel \#(0010) \parallel \#(000\bar{1})$
$= \{(\bar{1}000)\} \parallel \{(00\bar{1}0)\} \parallel \{(0001)\}$
$= \{(\bar{1}0\bar{1}1)\}$

Theorem 7.3.7 *Given $t_1 = t@x$ and $t_2 = t@y$, where x and y are sets of source vectors, if $x = \#(y)$ then $e(t_1) = \neg e(t_2)$, where $e(t_i)$ denotes the expression corresponding to t_i.* ■

Example 7.3.8 *Let $y = \{(1000), (0010), (000\bar{1})\}$ be a source vector. Then $x = \#(y) = \{(\bar{1}0\bar{1}1)\}$.*
Thus $e(t_2) = (f_1) \vee (f_3) \vee (\neg f_4)$. Further, $e(t_1) = (\neg f_1) \wedge (\neg f_3) \wedge (f_4) = \neg((f_1) \vee (f_3) \vee (\neg f_4)) = \neg e(t_2)$.

The UNION operation for source vectors is the same as the classical set theory operation. Let $x = \{u_1, ..., u_p\}$ and $y = \{v_1, ..., v_q\}$ be sets of source vectors, then $x \cup y = \{u_1, ..., u_p, v_1, ..., v_q\}$.

Example 7.3.9 *Let y, z be two source vectors as given below.*

$$y = \{(1000), (0010), (000\bar{1})\}$$

and

$$z = \{(1000), (00\bar{1}0), (000\bar{1})\}$$

Then $y \cup z = \{(1000), (0010), (000\bar{1}), (1000), (00\bar{1}0), (000\bar{1})\}$
$= \{(1000), (0010), (000\bar{1}), (00\bar{1}0)\}$.

Theorem 7.3.10 *Given $t_1 = t@x, t_2 = t@y$, and $t_3 = t@z$, where x, y, and z are sets of source vectors, if $x = y \cup z$ then $e(t_1) = e(t_2) \vee e(t_3)$, where $e(t_i)$ denotes the expression corresponding to t_i.* ■

Example 7.3.11 *Let y, z be as in Example 7.3.9. Then*
$e(t_1) = (f_1) \vee (f_3) \vee (\neg f_4) \vee (\neg f_3)$.
Now,
$e(t_2) = (f_1) \vee (f_3) \vee (\neg f_4)$
and
$e(t_3) = (f_1) \vee (\neg f_3) \vee (\neg f_4)$
Therefore,
$e(t_2) \vee e(t_3) = ((f_1) \vee (f_3) \vee (\neg f_4)) \vee ((f_1) \vee (\neg f_3) \vee (\neg f_4)) = (f_1) \vee (f_3) \vee (\neg f_4) \vee (\neg f_3) = e(t_1)$.

Extended relational algebra operations

Here we summarize extended relational algebra operations as defined in [35]. Interested reader is referred to [31, 32, 33, 34, 35] for a detailed discussion. It may be noted that the source attribute S is not visible to users and as such cannot be referenced in any query. This is another major difference that exists between IST and ciset relational database. Since source vector is not visible to the user and hence can not be used in any query, extended relational operations *selection* as well as *projection* do not differ from the (non-extended) selection and projection operation at all. Thus we have the following:

$$\sigma' c(r) = \{t@u \mid t@u \in r, \text{and } t \text{ satisfies condition } C\}$$

$$\Pi'_X(r) = \{t[X]@u \mid t@u \in r\}$$

The extended *union* is defined as in set union. Thus,

$$r \cup' s = \{t@u \mid t@u \in r \text{ or } t@u \in s\}$$

It is worth noticing that an implicit UNION operation takes place for source vectors in the above three extended operations.

We now proceed to define extended *intersection*, extended *Cartesian product*, and extended *natural join*. These three extended operations are defined using the 3OR operation for source vectors:

$$r \cap' s = \{t@(u_1 \parallel u_2) \mid t@u_1 \in r \text{ and } t@u_2 \in s\}$$
$$r \times' s = \{t_1 \cdot t_2(u_1 \parallel u_2) \mid t_1@u_1 \in r \text{ and } t_2@u_2 \in s\}$$
$$r \bowtie' s = \{t_1 \circ t_2(u_1 \parallel u_2) \mid t_1@u_1 \in r,\ t_2@u_2 \in s \text{ , and } t_1 \text{ and } t_2 \text{ join}\}$$

where $t_1 \cdot t_2$ indicates the concatenation of t_1 and t_2, and $t_1 \circ t_2$ indicates the join of t_1 and t_2. That is, $t_1 \circ t_2$ is the concatenation of t_1 and t_2 with the removal of duplicate values of common attributes.

The extended relational operation *difference* uses NEGATION and 3OR of source vectors and is defined as below.

$$r -' s = \{t@x \mid t@x \in r, \text{ and } t \text{ does not appear in } s, \text{ or } t@y \in r,\ t@z \in s, \text{ and } x = y \parallel (\#z)\}$$

Example 7.3.12 *Consider the relation food that lists foods and some of their ingredients; and dinner that lists the food served at dinner. "List people that will receive fish" can be computed using the extended relational operations as follows:*

FOOD

FDITEM	*INGDNT*	*I*
pasta	*vegetable*	1 0 0 0 0
pizza	*cheese*	1 0 0 0 0
pizza	*cheese*	0 1 0 0 0
pizza	*cheese*	0 0 0 0 1
pizza	*meat*	0 0 0 1 0
pizza	*meat*	0 0 1 0 0

DINNER

FDITEM	*PERSON*	*I*
pasta	*James Olsen*	1 0 0 0 0
pizza	*Liza Shue*	1 0 0 0 0
pizza	*Liza Shue*	0 1 0 0 0
pizza	*Liza Shue*	0 0 0 0 1
pizza	*Mike Cox*	0 1 0 0 0
pizza	*John Smith*	0 0 0 0 1

$\Pi_{PERSON}(DINNER \bowtie \sigma_{INGDT="meat"} FOOD)$

PERSON	*I*
Liza Shue	1 0 1 0 0
Liza Shue	0 1 1 0 0
Liza Shue	0 0 1 1 0
Liza Shue	1 0 0 1 0
Liza Shue	0 1 0 1 0
Liza Shue	0 0 0 1 1
Mike Cox	0 1 1 0 0
Mike Cox	0 1 0 1 0
John Smith	0 0 1 0 1
John Smith	0 0 0 1 1

TABLE 7.5 extended relations EMPLOYEE and STUDENT

EMPLOYEE

ID	ENAME	DEPT	I
14265	Mark Daub	Marketing	1 0 0
18945	Lee Wong	Mathematics	1 0 0
18945	Lee Wong	Mathematics	0 1 0
73912	Cathy Cox	Accounting	0 0 1

STUDENT

ID	ENAME	MAJOR	I
14265	Mark Daub	Marketing	0 1 0
23426	Kris Lamb	English	1 0 0
84909	Cory Work	Music	0 1 0
73912	Cathy Cox	Accounting	0 1 0

Example 7.3.13 *The answer to the query "Find all students who are not employees" is computed using extended relational operations as shown below.*

$STUDENT -' EMPLOYEE$

ID	*ENAME*	*MAJOR*	*I*
14265	*Mark Daub*	*Marketing*	$\bar{1}$ 1 0
23426	*Kris Lamb*	*English*	1 0 0
84909	*Cory Work*	*Music*	0 1 0
73912	*Cathy Cox*	*Accounting*	0 1 $\bar{1}$

The extended operations are precise

Now we proceed to prove that the extended relational operators are precise with respect or the representation discussed in this chapter.

Theorem 7.3.14 *The extended relational operations selection, projection, and union are precise.*

Proof. We will prove that the theorem for the extended union operation. The cases for selection and projection are similar. We shall show that, for all extended relations r and s,

$$rep(r \cup' s) = rep(r) \cup rep(s)$$

where $\cup'$ is the extended union operation.

Let S be the set of information sources, and $Q \subseteq S$. Let $p = r \cup' s$. We shall show that (1) if a tuple $t \in p^*(Q)$, then $t \in r^*(Q) \cup s^*(Q)$, and (2) if $t \in r^*(Q) \cup s^*(Q)$, then $t \in p^*(Q)$, where r^*, s^*, and p^* are the functions represented by extended relations r, s, and p, respectively.

(1) Assume that $t \in p^*(Q)$. Therefore, there exists a set of source vectors z such that $t@z \in p$ and $e(t@z) = true$ under truth assignment $truth(Q)$. Consequently, there exists a source vector $u \in z$ such that $e(t@u) = true$ under truth assignment $truth(Q)$. Now, $p = r \cup' s$. Therefore, either $t@u \in r$ or $t@u \in s$ or $t@u$ is in both r and s. It follows that $t \in r^*(Q)$ or $t \in s^*(Q)$ or t is in both $r^*(Q)$ and $s^*(Q)$. Thus $t \in r^*(Q) \cup s^*(Q)$.

(2) If $t \in r^*(Q) \cup s^*(Q)$ then $t \in r^*(Q)$ or $t \in s^*(Q)$. Hence there exists a set of source vectors z such that $t@z \in r$ or $t@z \in s$ and $e(t) = true$ under truth assignment $truth(Q)$. It follows that there is a set of source vectors $y, x \subseteq y$ and $t@y \in p$. Further, $e(t@y)$ is also $true$ under truth assignment $truth(Q)$ Note that adding disjuncts to a true expression cannot make it false. Consequently, $t \in p^*(Q)$. ■

Example 7.3.15 *In this example, we illustrate the fact that union is precise. Consider the extended relations FACULTY and FCLTY.*

FACULTY

F_ID	*FNAME*	*DEPT*	*I*
12312	*John Smith*	*Marketing*	100
31897	*Mary Lee*	*Mathematics*	100
31897	*Mary Lee*	*Mathematics*	010
56739	*Bea Anthony*	*Accounting*	001

FCLTY

F_ID	*FNAME*	*DEPT*	*I*
12312	*John Smith*	*Marketing*	010
31897	*Mary Lee*	*Mathematics*	100
31897	*Mary Lee*	*Mathematics*	001
56739	*James Jones*	*Accounting*	010

Now $FACULTY \cup' FCLTY$ is as follows:
$FACULTY \cup' FCLTY$

F_ID	FNAME	DEPT	I
12312	John Smith	Marketing	100
31897	Mary Lee	Mathematics	100
31897	Mary Lee	Mathematics	010
56739	Bea Anthony	Accounting	001
12312	John Smith	Marketing	010
31897	Mary Lee	Mathematics	001
56739	James Jones	Accounting	010

These relation in fact represents eight (regular) relations corresponding to eight different subsets of S. Let s_1, s_2, s_3. denote the three sources. Then there are eight subsets:
$\emptyset, \{s_1\}, \{s_2\}, \{s_3\}, \{s_1, s_2\}, \{s_2, s_3\}, \{s_1, s_3\}, S$.
We shall illustrate the case corresponding to the set $Q = \{s_1, s_2\}$. The other cases are similar.
$FACULTY^*(Q)$

F_ID	FNAME	DEPT
12312	John Smith	Marketing
31897	Mary Lee	Mathematics

$FCLTY^*(Q)$

F_ID	FNAME	DEPT
12312	John Smith	Marketing
31897	Mary Lee	Mathematics
56739	James Jones	Accounting

Now, $(FACULTY \cup' FCLTY)^*(Q)$ is given by
$(FACULTY \cup' FCLTY)^*(Q)$

F_ID	FNAME	DEPT
12312	John Smith	Marketing
31897	Mary Lee	Mathematics
56739	James Jones	Accounting

and is the same as $FACULTY^*(Q) \cup FCLTY^*(Q)$.

The following result is quite interesting as well as very useful in this context. On the other hand, it is quite obvious. Hence we state it without any proof.

Theorem 7.3.16 *Let S be the set of information sources, and $Q_1, Q_2 \subseteq S$. Then for all extended relations r, $r^*(Q_1 \cup Q_2) = r^*(Q_1) \cup r^*(Q_2)$.* ■

Proof of the following corollary is quite immediate and hence we state it without any proof.

Corollary 7.3.17 *Let $\otimes$ be a unary relational operation. A unary extended relational operation $\otimes'$ is precise, if and only if*

$$(\otimes'(r))^*(\{s\}) = \otimes(r^*(\{s\}))$$

for all extended relations r and any single source s. ∎

Corollary 7.3.18 *Let $\otimes$ be a binary relational operation. A binary extended relational operation $\otimes'$ is precise, if and only if*

$$(r \otimes' q)^*(\{s\}) = (r^* \otimes q^*)(\{s\}))$$

for all extended relations r and q; and any single source s. ∎

Theorem 7.3.19 *The extended relational operations intersection, Cartesian product, and natural join are precise.*

Proof. We will prove the theorem for Cartesian product. The proof of cases intersection and natural join are similar and hence omitted.

Let S be the set of information sources, and $Q \subseteq S$. Let $p = r \times' s$, where $\times'$ is the extended Cartesian product operation. We shall show that (1) if a tuple $t \in p^*(Q)$, then $t \in r^*(Q) \times s^*(Q)$, and (2) if $t \in r^*(Q) \times s^*(Q)$, then $t \in p^*(Q)$.

(1) If $t \in p^*(Q)$, then there exists a set of source vectors z such that $t@z \in p$ and $e(t@z) = true$ under the truth assignment $truth(Q)$. Hence, there exists a source vector $u \in z$ such that $e(t@u) = true$ under the truth assignment $truth(Q)$. Since $p = r \times' s$ it follows that there exist source vectors u_1 and u_2, and pure tuples t_1 and t_2 such that $t_1@u_1 \in r$, $t_2@u_2 \in s$, $t = t_1 \cdot t_2$, and $u = u_1 \parallel u_2$. By Theorem 7.3.3, $e(t@u) = e(t_1@u_1) \wedge e(t_2@u_2)$. It follows that $e(t_1@u_1) = e(t_2@u_2) = true$ under the truth assignment $truth(Q)$. Hence, $t_1 \in r^*(Q)$ and $t_2 \in s^*(Q)$. Then $t = t_1 \cdot t_2 \in r^*(Q) \times s^*(Q)$.

(2) If $t \in r^*(Q) \times s^*(Q)$ then there exists pure tuples $t_1 \in r^*(Q)$ and $t_2 \in s^*(Q)$, and $t = t_1 \cdot t_2$. It follows that there exist source vectors u_1 and u_2 such that $t_1@u_1 \in r$, $t_2@u_2 \in s$, and $e(t_1@u_1) = e(t_2@u_2) = true$ under the truth assignment $truth(Q)$. By the extended Cartesian operator, $t@u \in p$, where $t = t_1 \cdot t_2$ and $u = u_1 \parallel u_2$. Further, by Theorem 7.3.3, $e(t@u) = e(t_1@u_1) \wedge e(t_2@u_2) = true$ under the truth assignment $truth(Q)$. ∎

Theorem 7.3.20 *The extended set difference operation is precise.*

Proof. Let S be the set of information sources, and $Q \subseteq S$. Let $p = r -' s$, where $-'$ is the extended set difference operation. We shall show that (1) if a tuple $t \in p^*(Q)$, then $t \in r^*(Q) - s^*(Q)$, and (2) if $t \in r^*(Q) - s^*(Q)$, then $t \in p^*(Q)$.

(1) If $t \in p^*(Q)$, then there exists a set of source vectors z such that $t@z \in p$ and $e(t@z) = true$ under the truth assignment $truth(Q)$. Since $p = r -' s$, then two cases can happen.

(i) $t@z \in r$, and pure tuple t does not appear in s. It follows that $t \in r^*(Q)$ and t is not in $s^*(Q)$, hence $t \in r^*(Q) - s^*(Q)$.

(ii) There exists sets of source vectors y and z such that $t@y \in r$, $t@z \in s$, and $x = y \parallel (\#z)$. By Theorems 7.3.3 and 7.3.10 we have $e(t@x) = e(t@y) \wedge (\neg e(t@z))$. Hence, $e(t@y) = true$ and $e(t@z) = false$ under the truth assignment $truth(Q)$. It follows that $t \in r^*(Q) - s^*(Q)$.

(2) If $t \in r^*(Q) - s^*(Q)$ then either (i) t does not appear in s at all, or (ii) $t@z \in s$ for a set of source vectors z such that $e(t@z) = false$ under the truth assignment $truth(Q)$. In case (i) there exists a set of source vectors y such that $t@y \in r$ and $e(t@y) = true$ under the truth assignment $truth(Q)$. Hence, $t@y \in p$, and, as a result, $t \in p^*(Q)$. In case (ii) also there exists a set of source vectors y such that $t@y \in r$ and $e(t@y) = true$ under the truth assignment $truth(Q)$. But now $t@(y \parallel (\#z)) \in p$. However, $e(t@(y \parallel (\#z)) = e(t@y) \wedge (\neg e(t@z))$ and is $true$ under the truth assignment $truth(Q)$ by Theorems 7.3.3 and 7.3.10. It follows that $t \in p^*(Q)$. ■

Complexity of extended operations

In this section we compare the extended relational model with the classical model in the context of efficiency in query processing. The measure of interest is the time required for disk I/O operations which, for the majority of the applications, is the dominating cost factor. We will show that extended relational algebra operations have the same asymptotic I/O complexity as the classical relational algebra operations. However, the size of the database relations may be larger in the extended model, resulting in the degradation of query processing efficiency. This situation occurs mainly due to the fact that on the average there are more than one information source confirm the same piece of information in the database. To remedy the loss of query efficiency, we will present the Dual approach to Information Source Tracking in the next section. In the Dual approach the extended relations have the same number of tuples as the corresponding "pure" relations eliminating the main cause of loss of efficiency.

Size of extended relations

Let us first define the pure relation r^p corresponding to an extended relation r. Let $R = \{A_1, ..., A_n\}$ be an extended relation scheme, and r be a relation (instance) on R. The pure relation r^p corresponding to r is a non-extended (classical) relation consisting of the pure tuples in r, that is

$$r^p = \{t \mid t@u \in r\}$$

The pure relation r^p corresponding to an extended relation r contains the same data as r, except for information regarding the source(s) of the data.

The size of an extended relation r is larger than the corresponding pure relation r^p for two reasons:

1. **Tuple width.** Extended tuples have an additional attribute, the information source attribute I. A straightforward implementation using bit vectors increases the width of each tuple by $2k$ bits.
2. **Number of tuples.** A piece of information in the database may be confirmed by more than one information source. This means a pure tuple can be associated with more than one information source. This results in an increase in the "height" of an extended relation, compared to its corresponding pure relation.

To quantify the increase in the size of an extended relation, we define, for an extended relation r, the size increase factor $\bar{k}$ as

$$\bar{k}_r = \frac{size\ of\ r}{size\ of\ r^p}$$

Complexity of extended relational algebra operations

Selection, projection, and union

These operations are the same for the classical and extended models:

$\sigma'_c(r) = \{t@u \mid t@u \in r, \text{ and } t \text{ satisfies condition } C\}$
$\Pi'_X(r) = \{t[X]@u \mid t@u \in r\}$
$r \cup' s = \{t@u \mid t@u \in r \text{ or } t@u \in s\}$

Hence the complexities of the extended selection, projection, and union are the same as the complexities of their non-extended counterparts as a function of the size(s) of input relation(s). For example, a selection $\sigma'_c(r)$ on the extended relation r, carried out by a sequential search through the tuples of r (e.g. no index exists that can be used for this selection), will take $\bar{k}_r$ times the time for $\sigma_c(r^p)$ using the same sequential search.

In general, we may assume that complexities of these operations are linear in the size of the input relation(s) (Union operation needs a hashing scheme for duplicate elimination. Otherwise sorting is needed increasing the complexity if Union to $n \log n$. Performance of the extended model deteriorates by a factor of $\bar{k}_r$ for these operations due to the increase in the size of the extended relation (compared with pure operations carried out on corresponding pure relations).

Natural join

The extended natural join operation is defined as:

$$r \bowtie' s = \{(t_1 \circ t_2(u_1 \parallel u_2) \mid t_1@u_1 \in r, t_2@u_2 \in s, \text{ and } t_1 \text{ and } t_2 \text{ join})\}$$

Except for the 3OR operation between source vectors, the extended natural join operation is the same as the classical one. The 3OR being a computation and as such does not change the I/O complexity of the extended natural join.

The natural join can be performed in $n \log n$ time using a sorting scheme (e.g. sort the two relations on the joining attributes first, then join by scanning the relations sequentially in parallel). a more efficient algorithm which yields almost linear complexity under reasonable assumptions is the "Hybris Hash" scheme. Using such a scheme, $r \bowtie' s$ can be obtained for extended relations r and s in $\bar{k}$ times using the time it takes to compute $r^p \bowtie s^p$, where $\bar{k} = max\left(\bar{k}_r \times \bar{k}_s\right)$. We should also note that here that the size of $r \bowtie' s$ is, on the average, $\bar{k}_r \times \bar{k}_s$ times the size of $r^p \bowtie s^p$.

Cartesian product, intersection

These operations are less frequently needed in query formulation. Similar to the natural join, they use the 3OR operation on source vectors. As discussed above the CPU time does not change the I/O complexity of these operations.

$$r \cap' s = \{(t@(u_1 \parallel u_2) \mid t@u_1 \in r, \text{ and } t@u_2 \in s)\}$$
$$r \times' s = \{(t_1 \cdot t_2 (u_1 \parallel u_2) \mid t_1@u_1 \in r, \text{ and } t_2@u_2 \in s)\}$$

Similar to the natural join, the extended intersection operation can be performed using a hashing scheme with a near linear complexity. It follows that $r \cap' s$ can be performed in $\bar{k}$ times the time for $r^p \cap s^p$, where $\bar{k} = max\left(\bar{k}_r, \bar{k}_s\right)$.

The extended Cartesian product $r \times' s$ produces a relation which is $\bar{k}_r \times \bar{k}_s$ times the size of $r^p \times s^p$. Hence the extended Cartesian product is $\bar{k}_r \times \bar{k}_s$ times more costly than its pure counterpart. Note that our analysis of the complexity of extended natural join above makes the implicit assumption that the resulting relation size is comparable to the sizes of the input relations. In cases where the natural join degenerates to the Cartesian product, the complexity increases from $\bar{k}$, approaching $\bar{k}^2$ due to the sheer size of the output.

Set difference

Set difference uses NEGATION and 3OR of source vectors:

$r -' s = \{t@x \mid t@x \in r$, and the pure tuple t does not appear in s, or, $t@y \in r, t@z \in s$, and $x = y \parallel (\#z)\}$

The NEGATION and 3OR operations take up CPU time and do not change the I/O complexity. Similar to previous cases, a hashing scheme can be used to perform the set difference in linear time. Hence, the extended set difference $r -' s$ can be performed in $\bar{k}$ times the time for $r^p - s^p$.

We have demonstrated that although extended relational algebra operations have the same asymptotic complexity as the classical operations, yet query processing efficiency deteriorates due to the larger sizes of extended relations. The increase in query processing time is, for most queries, proportional to $\bar{k}$, where $\bar{k}$ is the average size increase factor of the extended relations in the database. For queries containing Cartesian product operations or degenerate natural joins, the query evaluation time may be increased by a factor of $\bar{k}^2$.

We envision for most database applications $\bar{k}$ will be close to unity, i.e. usually a piece of information is confirmed by a single information source. In these cases, the increase in query processing time will be small and tolerable.

For those applications where $\bar{k}$ is large, the increase in query processing time could become unacceptable. The "Dual" approach to Information Source Tracking, discussed later, is designed to provide the advantages of IST, yet use extended relation that have the same number of tuples as their pure counterparts.

7.4 Probabilistic Approach

Up to now, we had concentrated on the problem of determining information sources that contribute to an answer to a query. It is often possible to attach a quantitative reliability measure to each information source. In this section we will discuss how the reliability of an answer to a query can be calculated from the reliabilities of the information sources. We will further develop the alternate worlds model of Section 3, summarize reliability calculation algorithms from [31] and justify these algorithms.

There is a heated debate regarding the appropriate approach to the representation and management of uncertainty. The major approaches are the mathematical probability theory, certainty factors , and fuzzy sets. We have chosen the mathematical probability theory to quantify the reliabilities of the information sources. It should be emphasized that the identification of information sources contributing to an answer to a query is independent of the selection of the quantitative reliability measure.

Definition 7.4.1 *The probability of correctness of an information source* s_i *is called its reliability, and is denoted by* p_i.

Recall that we characterized the information content of an extended relation r as the function r^*, where, for a given subset Q of the set of information sources S,

$$r^*(Q) = \{t \mid t@x \in r,\ e(t@x) = true \text{ under } truth(Q)\}$$

With each $Q \subseteq S$, and subsequently with each $r^*(Q)$, we associate a probability $P(Q)$ as follows:

$$P(Q) = \prod_{s_i \in Q} p_i \prod_{s_i \in S-Q} (1 - p_i) \qquad (1)$$

In this way we have associated with each (regular) relation $r^*(Q)$ in the alternate world set of r a probability $P(Q)$ which is the probability that r represents $R^*(Q)$. In other words, not only we know r represents $r^*(Q)$, but we also have a quantitative measure of the likelihood that r represents $r^*(Q)$.

Example 7.4.2 *Consider the relation of Example 7.1.2. Assume the reliabilities of the three sources are as follows: $p_1 = 60\%$, $p_2 = 70\%$, and $p_3 = 80\%$. Then, the probabilities attached to the relations in the alternate world of the given relation, corresponding to the sets ϕ, $\{s_1\}$, $\{s_2\}$,$\{s_3\}$,$\{s_1, s_2\}$, $\{s_1, s_3\}$,$\{s_2, s_3\}$, and $\{s_1, s_2, s_3\}$ can be computed as follows. Let $Q = \varnothing$. Then $P(Q) = (1 - .6)(1 - .7)(1 - .8) = (.4)(.3)(.2) = .024$. That is, 2.4%. Now, let $Q = \{s_2, s_3\}$. Then $P(Q) = (1 - .6)(.7)(.8) = (.4)(.7)(.8) = .224$. In other words, $P(Q) = 22.4\%$ and so on.*

Given an extended relation r (e.g. representing the answer to a query), and a pure tuple t such that $t@x \in r$ for the set of source vector $x = \{u_1, ...u_p\}$, we would like to calculate the *reliability* of t. We define the reliability of t to be the degree to which r represents t. This, in turn, can be equated to the probability that t is present in the alternate world of r. We will make this notion precise below.

Definition 7.4.3 *The reliability of a pure tuple t, represented by a relation r (i.e. $t@x \in r$, for a set of source vectors $x = \{\{u_1, ...u_p\}\}$), is*

$$re(t) = \sum_{t \in r^*(Q)} P(Q) \qquad (2)$$

First, let us assume that $x = \{u\}$, e.g. t has a single source vector associated with it in r. Let $u = (a_1...a_k)$. We can partition the set of sources I into three sets: $S^+(u)$, $S^-(u)$, $S^0(u)$ as follows:

$$S^+(u) = \{s_i \mid a_i = 1\}$$
$$S^-(u) = \{s_i \mid a_i = \bar{1}\}$$
$$S^0(u) = \{s_i \mid a_i = 0\}$$

Intuitively, $S^+(u)$ is the set of information sources contributing positively to a tuple $t@u$; $S^-(u)$ is the set of information sources contributing negatively to $t@u$; and $S^0(u)$ is the set of information sources not contributing to $t@u$.

Lemma 7.4.4 *If $t@u \in r$, then the pure tuple t appears in (regular) relations $r^*(Q)$ from the alternate world of r if and only if $S^+(u) \subseteq Q \subseteq S^+(u) \cup S^0(u)$.*

Proof. Recall that

$$e(t@u) = \bigwedge_{s_i \in S^+} f_i \bigwedge_{s_i \in S^-} \neg f_i$$

Obviously $e(t@u) = true$ under $truth(Q)$ if and only if $S^+(u) \subseteq Q \subseteq S^+(u) \cup S^0(u)$. ∎

Theorem 7.4.5 *If $t@u \in r$ ($t@u$ is the only (extended) tuple in r corresponding to the pure tuple t), then*

$$re(t) = \prod_{s_i \in S^+(u)} p_i \prod_{s_i \in S^-(u)} (1 - p_i) \tag{3}$$

where $re(t)$ is the reliability of t.

Proof. By the definition of the reliability of a tuple (Equation 2), and Lemma 7.4.4,

$$re(t) = \sum_{S^+(u) \subseteq Q \subseteq S^+(u) \cup S^0(u)} p(Q) \tag{4}$$

It is a simple exercise to replace $P(Q)$ by its expression from Equation1, and reduce the above summation to obtain

$$re(t) = \prod_{s_i \in S^+(u)} p_i \prod_{s_i \in S^-(u)} (1 - p_i)$$

∎

We shall now concentrate on the general case, where $t@x \in r$ where x has more than one source vector in it. In [35] two algorithms were presented for the calculation of the reliability of t. We will use Equation 2, which is based on the alternate world model, to justify these algorithms.

Lemma 7.4.6 *Assume $t@x \in r$, $x = \{u_1, ..., u_p\}$. Then $t \in r^*(Q)$ if and only if $S^+(u_i) \subseteq Q \subseteq S^+(u_i) \cup S^0(u_i)$, for at least one $i, 1 \leq i \leq p$.*

Proof. Obvious from Lemma 7.4.4, and the fact that

$$e(t@x) = \bigvee_{i=1}^{p} e(t@u_i)$$

■

Now we can derive an expression, similar to equation 4, for the reliability of a tuple in the general case (i.e. $t@x \in r$ where $x = \{u_1, ..., u_p\}$). Let

$$Q = \{Q_i \mid S^+(u_i) \subseteq Q \subseteq S^+(u_i) \cup S^0(u_i), i = 1, ..., p\} \tag{5}$$

then

$$re(t) = \sum_{Q \in Q} P(Q) \tag{6}$$

It is very important to notice that in general

$$re(t@x) \neq re(t@u_1) + ... + re(t@u_p)$$

where $re(t@x)$ is the reliability of t when $t@x \in r$, and $re(t@u_i)$ is the reliability of t when *only* $t@u_i \in r$. This os because some of $P(Q)$'s in Equation 6 may be added more than once when we sum up $re(t@u_i)$'s. This observation led to the reliability calculation algorithms of [31] which we summarize below.

Reliability calculation algorithms

ALGORITHM 1: Assume $t@x \in r$, $x = \{u_1, ..., u_p\}$. Let

$$K_1 = \Sigma_{i=1}^{p} re(u_i)$$
$$K_2 = \Sigma_{i=1}^{p} \Sigma_{j \succ i}^{p} re(u_i \parallel u_j)$$
$$K_3 = \Sigma_{i=1}^{p} \Sigma_{j \succ i}^{p} \Sigma_{k \succ j}^{p} re(u_i \parallel u_j \parallel u_k)$$
$$...K_p = re(u_i \parallel u_j)$$

Then

$$re(t) = K_1 - K_2 + K_3 - ... + (\bar{1})^{p-1} K_p \tag{7}$$

ALGORITHM 2: Consider the expression for $t@x$

$$e(t@x) = \bigvee_{i=1}^{p} e(t@u_i)$$

This expression is in disjunctive form, (also called "sum-of-products") where each $e(t@u_i)$ in a conjuct. Convert this expression into *disjunctive normal form*, i.e.

$$e(t@x) = \bigvee_{i=1}^{p} e(t@v_i)$$

where each $e(t@v_i)$ is a conjuct in which all variables $f_1, ..., f_k$ (maybe negated) appear. Note that the disjunctive normal form of a Boolean expression is unique (up to a permutation of conjucts). Then

$$re(t) = re(t@v_1) + ... + re(t@v_q) \tag{8}$$

Example 7.4.7 *Assume $t@\{(10\bar{1}), (110)\} \in r$. Let the reliability of information source be 60%, 70%, and 80%, respectively. We can calculate the reliability of t using both algorithms given above. Applying Algorithm 1:*

$K_1 = re(t@(10\bar{1})) + re(t@(110)) = (.6)(1-.8) + (.6)(.7) = .12 + .42 = .54$
$K_2 = re(t@(10\bar{1} \parallel 110)) = re(t@(11\bar{1})) = (.6)(.7)(.2) = .084$
$re(t) = K_1 - K_2 = .54 - .084 = .456 = 45.6\%$

Now,to use Algorithm 2, first we convert $e(t)$ to disjunctive normal form as follows.

$e(t) = (f_1 \wedge \neg f_3) \vee (f_1 \wedge f_2)$
$e(t) = (f_1 \wedge f_2 \wedge f_3) \vee (f_1 \wedge f_2 \wedge \neg f_3) \vee (f_1 \wedge \neg f_2 \wedge \neg f_3)$
producing the extended tuple $t@(\{(111), (11\bar{1}), (1\bar{1}\bar{1})\})$. Consequently,
$re(t) = re(\ t@(111)) + re(t@(11\bar{1})) + re(t@(1\bar{1}\bar{1}))$
$re(t) = (.6)(.7)(.8) + (.6)(.7)(1 - .8) + (.6)(1 - .7)(1 - .8)$
$re(t) = .336 + .084 + .036 = .456 = 45.6\%$

Theorem 7.4.8 *Algorithm 1 correctly computes the reliability of a pure tuple t given by the formula $re(t) = \sum_{t \in r^*(Q)} P(Q)$.*

■

Lemma 7.4.9 *Let $Q_i = \{Q_i \mid S^+(u_i) \subseteq Q_i \subseteq S^+(u_i) \cup S^0(u_i)\}, i = 1, 2.$*

Then

$$\sum_{Q \in Q_1 \cup Q_2} P(Q) = \sum_{Q \in Q_1} P(Q) + \sum_{Q \in Q_2} P(Q) - \sum_{Q \in Q_1 \cap Q_2} P(Q)$$

Proof. This lemma is based on the principle of inclusion and exclusion. Some of the Q's may be repeated in the first and second summation on the right hand side, which are deducted in the third summation on the right hand side.

This lemma can be generalized to more than two sets. The generalization will alternate adding and subtracting the effect appearing in only one set, in two sets, in three sets, etc. We omit the details here. ■

Proof of Theorem 7.4.8: We obtained Equation 6 for the reliability of t using Lemma 7.4.6, so we will show that the algorithm computes $re(t)$ according to Equation 6.

In the light of Lemma 7.4.9, all we need to show is the following: Let $\mathcal{Q}_i = \{Q_i \mid S^+(u_i) \subseteq Q_i \subseteq S^+(u_i) \cup S^0(u_i)\}, i = 1, 2$. Then $\mathcal{Q} = \mathcal{Q}_1 \cap \mathcal{Q}_2$ if and only if $\mathcal{Q} = \{Q \mid S^+(u) \subseteq Q \subseteq S^+(u) \cup S^0(u)\}$ where $u = u_1 \parallel u_2$. This follows from Theorem 7.3.3. Since $e(t@u_1) = true$ under $truth(Qi)$ if and only if $Q_i \in \mathcal{Q}_i$, and $e(t@u) = e(t@u_1) \wedge e(t@u_2)$ is true if and only if both $e(t@u_1) = e(t@u_2) = true$, if and only if $Q \in \mathcal{Q}_1 \cap \mathcal{Q}_2$. The generalization to the case of more than two source vectors is straightforward and we omit it here. ■

Theorem 7.4.10 *Algorithm 2 correctly computes the reliability of a pure tuple t according to Equation 2.*

Proof. The observation to make here is if $e(t@v_i)$ is a conjuct in which all variables $f_1, ..., f_k$ (maybe negated) appear, then $S^0(v_i) = \phi$ and thee set $\mathcal{Q}_i = \{Q_i \mid S^+(v_i) \subseteq Q_i \subseteq S^+(v_i) \cup S^0(v_i)\}$ is a singleton. in other words, once converted to disjunctive normal form, each $t@v_i$ makes $t \in r^*(Q)$ for exactly on Q. Since there are no repetitions in disjunctive normal form, Equation 8 correctly computes the reliability of t as specified by Equation 6. ■

The Dual IST model

As we seen in previous sections the efficiency of query processing deteriorates very rapidly in a Information Source Tracking system as number of information sources confirming the same information in the database increases. In this section we present an alternate representation proposed by Sadri [31, 32, 33, 34, 35]. The main idea is to represent the data with the least number of tuples, and thus eliminate the main cause of loss of efficiency of the system. To begin with, the same definition for the extended relational model is maintained, except the semantics interpretation of the information source vector is now different. The extended relational algebra operations have to be modified also to reflect the new sematic interpretation. We refer to this model as the dual IST model as there are similarities as in the case of duality principle in Boolean algebra.

An extended relation scheme is still defined as a set of attributes $\{A_1, ..., A_n, I\}$, where $A_1, ..., A_n$ are regular attributes, and I is the information source attribute. The domains of A_i and I are defined as before, and a tuple on the (extended) scheme $R = \{A_1, ..., A_n, I\}$ is an element of $D_1 \times ... \times D_n \times D_I$. However, the interpretation of source vector values are different for the Dual approach.

Let $t@x$ be a tuple in the extended relation r, where x is a set of source vectors $x = \{u_1, ..., u_p\}$. In the Dual approach, x is treated as the conjunction of the disjunction of its non-zero elements. Hence the expression corresponding to t in $t@x$ is treated as in conjuctive form or product of sums form. We now proceed to formalize these concepts.

First, consider the case where there is a single (extended) tuple $t@u$ in r, where $u = (a_1 ... a_k)$. We denote the set of information sources by $S = \{s_1, ..., s_k\}$, where k is the number of information sources. The sources $S^+ = \{s_i \mid a_i = 1\}$ are contributing positively to a tuple t, while the sources $S^- = \{s_i \mid a_i = \bar{1}\}$ are contributing negatively. We also associate with each information source s_i a Boolean variable f_i. The expression corresponding to t with respect to u, denoted $e\,(t@u)$, is defined as

$$e\,(t@u) = \bigvee_{s_i \in S^+} f_i \bigvee_{s_i \in S^-} \neg f_i$$

The expression corresponding to a zero source vector is the logical expression $True$, that is, $e\,(t@u) = True$ for $u = 0$. When $t@x \in r$, where $x = \{u_1, ..., u_p\}$, the expression corresponding to t with respect to x, is written as

$$e\,(t) = e\,(t@x) = \bigwedge_{i=1}^{p} e\,(t@u_i)$$

As before, we regard the expression corresponding to a tuple t in a relation r as a propositional logic expression, where $f_i, ..., f_k$ represent Boolean variables. A truth assignment $f_i = true$ is interpreted as "information source s_i is correct", otherwise, $f_i = false$ which indicates whether t is a valid tuple $(e\,(t) = true)$, or an invalid tuple $(e\,(t) = false)$.

Normally, information in a database is stored in relations (called base relations), where each tuple is confirmed by one or more information sources. The Dual representation makes it possible to have a single (extended) tuple representing a pure tuple with all its confirming information sources. For example, if sources s_1 and s_2 confirm that Mary Lee has 31897 as faculty identification number and is a faculty member of Department of Mathematics, the tuple (31897, Mary Lee, Mathematics, 110) is used in the Dual representation (assuming there are 3 information sources, that is, $k = 3$). Note that in the original IST scheme, the same information would require 2 extended tuples to represent.

It follows that there is no "vertical" size increase, in other words, the number of tuples in an extended relation r is the same as the number of tuples in its corresponding pure relation r^p. This would effectively decrease $\bar{k}$, the average size to increase, to close to unity. Of course, we need new algorithms for extended relational algebra operations, which will be discussed in the next section.

Example 7.4.11 *The relation faculty in Example 7.1.2 will be as given below in the Dual IST model. The expression corresponding to pure tuple (12312, John Smith, Marketing) is f_1. Similarly, the expression corresponding to the pure tuple (31897, Mary Lee, Mathematics) is given by $f_1 \vee f_2$. Further, expression corresponding to pure tuple (56739, Bea Anthony,Accounting) is f_3. Note that the original IST representation needed 4 tuples for the same information.*

TABLE 7.6 extended relation FACULTY in Dual IST model

FTY

F_ID	FNAME	DEPT	I
12312	John Smith	Marketing	100
31897	Mary Lee	Mathematics	110
56739	Bea Anthony	Accounting	001

Since the expression corresponding to FACULTY and FTY are identical for reach of the pure tuples involved, it follows that alternate worlds of FACULTY and FTY are also identical.

Source vector operations for the dual approach

The source vector operations NEGATION, and UNION are defined as before. We redefine 3OR. The effect of these operations corresponding to the new source vector interpretation in the Dual IST model is discussed below.

3OR operation on source vectors

We redefine the 3OR binary operation on source vectors. The 3OR of two non-zero source vectors, $u = (a_1 ... a_k)$ and $v = (b_1 ... b_k)$ is a source vector $w = (c_1 ... c_k)$ obtained using the following look up table:

a_i	b_i	c_i
$\bar{1}$	$\bar{1}$	$\bar{1}$
$\bar{1}$	0	$\bar{1}$
$\bar{1}$	1	0
0	$\bar{1}$	$\bar{1}$
0	0	0
0	1	1
1	$\bar{1}$	0
1	0	1
1	1	1

We write $w = u \parallel v$ to indicate the 3OR operation. Note that the main difference is $1 \parallel \bar{1}$ do not result in $w = 0$, but rather $1 \parallel \bar{1} = 0$. Let $u = (01\bar{1}0)$ and $v = (\bar{1}0\bar{1}1)$ then $u \parallel v = (\bar{1}1\bar{1}1)$. Note that if we restrict the values of a_i and b_i in the table to Boolean values, 0 and 1, then we obtain the following:

a_i	b_i	c_i
0	0	0
0	1	1
1	0	1
1	1	1

This is the Boolean OR operation where 0 represents false and 1 represents true. Similarly, if we restrict the values of a_i and b_i in the table to Boolean values, 0 and $\bar{1}$, then we obtain the following:

a_i	b_i	c_i
$\bar{1}$	$\bar{1}$	$\bar{1}$
$\bar{1}$	0	$\bar{1}$
0	$\bar{1}$	$\bar{1}$
0	0	0

This is once again the Boolean OR operation where 0 represents false and $\bar{1}$ represents true. Thus 3OR is in fact an extension of our traditional Boolean OR; and hence the name 3OR.

Thus both 1 and $\bar{1}$ behave identically as far as 3OR is concerned. This is an important feature of 3OR.

Theorem 7.4.12 *Given $t_1 = t@x, t_2 = t@y$, and $t_3 = t@z$, where x, y, and z are sets of source vectors, if $x = y \parallel z$ then $e(t_1) = e(t_2) \vee e(t_3)$, where $e(t_i)$ denotes the expression corresponding to t_i.*

Proof. First let us consider the case where $x = \{u\}$ and $y = \{v\}$, i.e. x and y are single source vectors. The "truth" table of the 3OR operation effectively implements the following Boolean algebra rules:

$$f_i \vee f_i = f_i$$
$$\neg f_i \vee \neg f_i = \neg f_i$$
$$f_i \vee \neg f_i = \neg f_i \vee f_i = True$$

When $x = \{u_1, ..., u_p\}$ and $y = \{v_1, ..., v_q\}$ the definition of the 3OR operation for sets of source vectors implements the Boolean algebra OR. ■

Example 7.4.13 *Let $y = \{(101\bar{1})\}$ and $z = \{(10\bar{1}\bar{1})\}$ be two source vectors. Then $y \parallel z = \{(100\bar{1})\}$*

$e(t_1) = (f_1) \vee (\neg f_4)$.

Now,

$e(t_2) = (f_1) \vee (f_3) \vee (\neg f_4)$

and

$e(t_3) = (f_1) \vee (\neg f_3) \vee (\neg f_4)$

Therefore,

$$\begin{aligned} e(t_2) \vee e(t_3) &= ((f_1) \vee (f_3) \vee (\neg f_4)) \vee ((f_1) \vee (\neg f_3) \vee (\neg f_4)) \\ &= (f_1) \vee (\neg f_4) \\ &= e(t_1). \end{aligned}$$

Example 7.4.14 *Let $y = \{(110\bar{1})\}$ and $z = \{(10\bar{1}\bar{1})\}$ be two source vectors. Then $y \parallel z = \{(11\bar{1}\bar{1})\}$*

$e(t_1) = (f_1) \vee (\neg f_4)$.

Now,

$e(t_2) = (f_1) \vee (f_2) \vee (\neg f_4)$

and
$e(t_3) = (f_1) \vee (\neg f_3) \vee (\neg f_4)$
Therefore,
$$e(t_2) \vee e(t_3) = ((f_1) \vee (f_2) \vee (\neg f_4)) \vee ((f_1) \vee (\neg f_3) \vee (\neg f_4))$$
$$= (f_1) \vee (f_2) \vee (\neg f_3)(\neg f_4)$$
$$= e(t_1).$$

Example 7.4.15 *Let* $y = \{(101\bar{1})\}$ *be a source vector. Then*
$\#(y) = \{(\bar{1}000), (00\bar{1}0), (0001)\}$

Example 7.4.16 *Let* $y = \{(101\bar{1})\}$ *be a source vector. Then* $x = \#(y) = \{(\bar{1}000), (00\bar{1}0), (0001)\}$.

Thus $e(t_2) = (f_1) \vee (f_3) \vee (\neg f_4)$. *Further,* $\neg e(t_2) = \neg((f_1) \vee (f_3) \vee (\neg f_4)) = (\neg f_1) \wedge (\neg f_3) \wedge (f_4) = e(t_1)$.

Example 7.4.17 *Let* $y = \{(101\bar{1}), (1\bar{1}0\bar{1})\}$ *be a set of source vectors. Then* $x = \#(y) = \{(\bar{1}000), (00\bar{1}0), (0001)\} \parallel \{(\bar{1}000), (0100), (0001)\}$
$= \{(\bar{1}000), (\bar{1}100), (\bar{1}001), (\bar{1}0\bar{1}0), (01\bar{1}0), (00\bar{1}1), (\bar{1}001), (0101), (0001)\}$

Thus $e(t_2) = ((f_1) \vee (f_3) \vee (\neg f_4)) \wedge ((f_1) \vee (\neg f_2) \vee (\neg f_4)) = .$ *Further,*
$\neg e(t_2) = \neg((f_1) \vee (f_3) \vee (\neg f_4)) \wedge ((f_1) \vee (\neg f_2) \vee (\neg f_4))$
$= ((\neg f_1) \wedge (\neg f_3) \wedge (f_4)) \vee ((\neg f_1) \wedge (f_2) \wedge (f_4))$
$= (\neg f_1) \wedge (\neg f_1 \vee f_2) \wedge (\neg f_1 \vee f_4) \wedge (\neg f_1 \vee \neg f_3)$
$\wedge (f_2 \vee \neg f_3) \wedge (\neg f_3 \vee f_4) \wedge (\neg f_1 \vee f_4) \wedge (f_2 \vee f_4) \wedge (f_4)$
$= e(t_1)$.

Theorem 7.4.18 *Given* $t_1@x$ *and* $t_2@y$, *where* x *and* y *are sets of source vectors, if* $x = \#(y)$ *then* $e(t_1) = \neg e(t_2)$, *where* $e(t)$ *denotes the expression corresponding to* t.

Proof. First let y be a single source vector, $y = \{v\}$, where $v = (a_1 ... a_k)$. Let $S^+ = \{s_i \mid a_i = 1\}$ and $S^- = \{s_i \parallel a_i = \bar{1}\}$, then

$$e(t_2@y) = \bigvee_{s_i \in S^+} f_i \bigvee_{s_i \in S^-} \neg f_i$$

and

$$\neg e(t_2@y) = \bigwedge_{s_i \in S^+} \neg f_i \bigwedge_{s_i \in S^-} f_i$$

this is exactly the expression corresponding to $t_1@x$ where $x = \#(v)$.

When $y = \{v_1, ..., v_q\}$, the definition of the source vector negation operation for a set of source vectors implements the Boolean algebra DeMorgan's law

$$\neg e(e_1 \wedge ... \wedge e_n) = \neg e_1 \vee ... \vee \neg e_n$$

■

Theorem 7.4.19 *Given $t_1@x$, $t_2@y$, and $t_3@z$, where x, y, and z are sets of source vectors, if $x = y \cup z$ then $e(t_1) = e(t_2) \wedge e(t_3)$, where $e(t)$ denotes the expression corresponding to t.*

Proof. Directly follows from the semantic interpretation of source vectors. That is, a set of source vectors represent the conjunction of the expressions corresponding to each of them. ■

Extended operations for the dual IST

Selection operation is performed as before:

$$\sigma_c(r) = \{t@u \mid t@u \in r, \text{ and } t \text{ satisfies condition } C\}$$

Note that condition C can not refer to the information source attribute I, which is not visible to users.

Union operation is more complicated:

$$\begin{aligned} r \cup s = \{t@z \mid\ & t@z \in r \text{ and } t \text{ does not appear in } s, \text{ or} \\ & t@z \in s \text{ and } t \text{ does not appear in } s, \text{ or} \\ & t@x \in r,\ t@y \in r \text{ and } z = x \parallel y\} \end{aligned}$$

We can justify the above formula by observing that $t@x \in r$ means that t is a valid tuple of r^p if $e(t@x)$ is true. Similarly t is a valid tuple s^p if $e(t@y)$ is true. It follows that t is a valid tuple in the union of r and s if $e(t@x) \vee e(t@y)$ is true. That is exactly what $z = x \parallel y$ achieves. The case where t appears only in one relation is straightforward.

The projection operation had some similarities with the union. There is the possibility that several pure tuples map to the same pure tuple after projection. The expressions for the resulting tuple is the disjunction of the expressions of the original tuples, which calls for the 3OR source vector operation:

$$\Pi_x(r) = \\ \{t@z \mid t_1@y_1, ..., t_n@y_n \in r, t_1[X] = ... = t_n[X] = t, \text{ and } z = y_1 \| ... \| y_n\}$$

Extended intersection, Cartesian product, and natural join use the source vector operation UNION:

$$\begin{aligned} r \cap s &= \{t@z \mid t@x \in r, t@y \in s, \text{ and } z = x \cup y\} \\ r \times s &= \{t_1 \cdot t_2@z \mid t_1@x \in r, t_2@y \in s, \text{ and } z = x \cup y\} \\ r \bowtie s &= \{t_1 \circ t_2@z \mid t_1@x \in r, t_2@y \in s,\ z = x \cup y, \text{ and } t_1 \text{ and } t_2 \text{ join}\} \end{aligned}$$

where $t_1 \cdot t_2$ indicates the concatenation of t_1 and t_2, and $t_1 \circ t_2$ indicates the join of t_1 and t_2, i.e. the concatenation of t_1 and t_2 with the removal of duplicate values of common attributes. Two tuples t_1 and t_2 join if they have the same values for the common attributes.

To justify the above operation we observe, for example, that a tuple t is valid in $r \cap s$ if t is in both r and s. This happens when both $e(t@x)$ and $e(t@y)$ are true, which is equivalent to $e(t@x) \wedge e(t@y)$ being true. Note that the source vector operation UNION implements logical conjunction, that is, $e(t@z) = e(t@x) \wedge e(t@y)$ for $z = x \cup y$.

Example 7.4.20 *In this example, we evaluate the query of Example 7.3.12 using the Dual representation. Consider relations food that lists foods and some of their ingredients; and dinner that lists the food served at dinner. "List people that will receive fish" can be computed using the extended relational operations as follows:*

FOOD

FDITEM	*INGDNT*	*I*
pasta	*vegetable*	1 0 0 0 0
pizza	*cheese*	1 1 0 0 1
pizza	*meat*	0 0 1 1 0

DINNER

FDITEM	*PERSON*	*I*
pasta	*James Olsen*	1 0 0 0 0
pizza	*Liza Shue*	1 1 0 0 1
pizza	*Mike Cox*	0 1 0 0 0
pizza	*John Smith*	0 0 0 0 1

$\Pi_{PERSON}(DINNER \bowtie \sigma_{INGDT="meat"} FOOD)$

PERSON	*I*
Liza Shue	1 1 0 0 1
Liza Shue	0 0 1 1 0
Mike Cox	0 1 0 0 0
Mike Cox	0 0 1 1 0
John Smith	0 0 0 0 1
John Smith	0 0 1 1 0

Finally, set differences uses NEGATION and UNION of source vectors:

$r - s = \{t@x \mid t@x \in r$, and the pure tuple t does not appear in s,

or,

$t@y \in r, t@z \in s$, and $x = y \cup (\#z)\}$

Example 7.4.21 *In this example, we evaluate the query of Example 7.3.13 using the Dual representation.*

TABLE 7.7 extended relations EMPLOYEE and STUDENT

EMPLOYEE

ID	ENAME	DEPT	I
14265	Mark Daub	Marketing	0 1 0
18945	Lee Wong	Mathematics	1 0 0

STUDENT

ID	ENAME	MAJOR	I
14265	Mark Daub	Marketing	1 0 0
23426	Kris Lamb	English	0 0 1

The answer to the query "Find all students who are not employees" is computed using extended relational operations is shown below.

STUDENT−′EMPLOYEE

ID	*ENAME*	*MAJOR*	*I*
14265	*Mark Daub*	*Marketing*	1 0 0
14265	*Mark Daub*	*Marketing*	0 $\bar{1}$ 0
23426	*Kris Lamb*	*English*	0 0 1

Complexity of dual extended operations

The motivation behind the Dual approach had been to reduce the size of the database (base) relations. Normally database information, presented by tuples in base relations, are confirmed by information sources independently. In the Dual approach one follows that the number of tuples can represent the data plus all confirming information sources. It follows that the number of tuples in an extended relation r in the Dual approach is equal to the number of tuples in the corresponding pure relation r^p. Hence the only size increase for base relations is due to the increase in tuple width. If the source vector values are represented by a $2k$ bit vector (k is the number of information sources), then:

$$size\ of\ r = \frac{(2k+1)}{l}\,(size\ of\ r^p)$$

where l is the pure tuple length (bits).
In most applications $l >> k$, and hence, for such applications:

$$\overline{k} = \frac{size\ of\ r}{size\ of\ r^p} \approx 1$$

It should be noted that some of the relational algebra operations, in the Dual approach result in an increase in the size of the result (as compared with the result of same operation performed in pure relations). The same situation existed with the original Information Source Tracking approach also. The Dual approach is advantageous even with this respect. We will see that in most cases this increase in size is smaller in the Dual approach as compared with the original IST.

In what follows we will compare the complexity of the extended relational algebra operations with the classical operations. As before, we concentrate in the I/O complexity of the operations.

Selection, union, projection

Extended selection obviously had the same complexity as the pure selection. For extended union we have

$$r \cup s = \{t@z \mid t@z \in r \text{ and } t \text{ does not appear in } s, \text{ or } t@z \in s \text{ and } t \text{ does not appear in } s, \text{ or } t@x \in r,\ t@y \in r \text{ and } z = x \parallel y\}$$

A mechanism is needed to group together extended tuples that have the same pure component. This can be achieved by sorting, or, to achieve linear time complexity, by hashing. The complexity of extended union is the same as the complexity if its pure counterpart implemented by the same technique (sorting or hashing). Note that if r and s are base relations, then the number of tuples in $r \cup s$ is equal to that of $r^p \cup s^p$: the 3OR of two (single) source vectors is a (single) source vector.

Similar to extended union, extended projection also has the same complexity as the pure projection. The number of pure tuples in the result, however, can be larger than the projection of the corresponding pure relation as a result of 3OR operation among sets of source vectors. This is the only case that the Dual approach is disadvantageous compared with the original IST.

$$\Pi_x(r) = \{t@z \mid t_1@y_1, ..., t_n@y_n \in r, t_1[X] = ... = t_n[X] = t, \text{ and } z = y_1 \| ... \| y_n\}$$

Intersection, Cartesian Product, Natural Join

The complexity of these operations is similar to those of the pure operations using the same method (sorting or hashing). The size of the result will be larger than the same operation carried out on the pure relations due to the source vector UNION operation. This increase, however, is additive for the Dual approach (as compared with the original IST where the increase is multiplicative). For example, if $t_1@x \in r$ and $t_2@y \in s$ join, then they produce $|x| + |y|$ tuples in the Dual approach, while $|x| \times |y|$ tuples are generated in the original IST ($|A|$ indicates the number of elements in the set A). We can verify this fact by observing that the operations intersection, Cartesian product, and natural join utilize 3OR operation in the original IST, while they utilize source vector UNION operation in the Dual approach:

$$r \cap s = \{t@z \mid t@x \in r, t@y \in s, \text{ and } z = x \cup y\}$$
$$r \times s = \{t_1 \cdot t_2@z \mid t_1@x \in r, t_2@y \in s, \text{ and } z = x \cup y\}$$
$$r \bowtie s = \{t_1 \circ t_2@z \mid t_1@x \in r, t_2@y \in s, \; z = x \cup y, \text{ and } t_1 \text{ and } t_2 \text{ join}\}$$

Set Difference

The last operation to discuss, extended set difference, is implemented using source vector operations UNION and NEGATION. The I/O time complexity of extended set difference is the same as the pure operation, the NEGATION is a CPU operation, and UNION is achieved by writing out the tuples. The size of the result may be larger than the size of set difference of the pure relations, while the Dual approach achieves an additive size increase (similar to previous cases) as compared to the original IST method that has a multiplicative size increase.

$r - s = \{t@x \mid t@x \in r$, and the pure tuple t does not appear in s,

or,

$t@y \in r, t@z \in s$, and $x = y \cup (\#z)\}$

All relational algebra operations studied have the same complexities, as a function of the size of input relation, for both the IST and the Dual approaches. The important difference between the two approaches is the in the IST approach the size of the base relations may become substantially larger than the size of their corresponding pure relations, while the Dual approach eliminates the major cause of size increase in the base relations, namely the increase in the number of tuples. A second advantage of the Dual approach is to achieve a slower (i.e. additive) size increase when operations are carried out versus the IST which has a multiplicative size increase.

Probability calculation algorithms for dual IST

Once the answer to a query is obtained using Dual IST, the reliability of the tuples in the answer can be calculated as a function of the reliabilities of information sources in a manner similar to that of IST. We present two algorithms in this section.

ALGORITHM 3: Assume $t@x \in r$, $x = \{u_1, ..., u_p\}$, where r is an extended relation in the Dual IST format. The expression for t is in conjuctive form

$$e(t) = e(t@x) = \bigwedge_{i=1}^{p} e(t@u_i)$$

where each $e(t@u_i)$ is a disjunct of terms of the form

$$e(t@u) = \bigvee_{s_i \in S^+} f_i \bigvee_{s_i \in S^=} \neg f_i$$

Convert $e(t)$ to disjunctive form, then use Algorithm 1 or 2 to calculate the reliability.

Example 7.4.22 *Assume* $t@\{(100), (01\bar{1})\} \in r$. *Let the reliability of information sources be* 60%, 70% *and* 80% *respectively. The expression for* t

$e(t) = f_1 \wedge (f_2 \vee \neg f_3)$

is converted to disjunctive form as

$e(t) = (f_1 \wedge f_2) \vee (f_1 \wedge \neg f_3)$

The reliability $re(t)$ *for this case was calculated in Example 7.4.7 and equal to* 45.6%.

ALGORITHM 4: Obtain the expression for $e'(t) = \neg e(t)$ using De-Morgan's rule, that is,

$$e'(t) = \neg e(t) = \bigwedge_{i=1}^{p} \neg e(t@u_i)$$

$e'(t)$ will be in disjunctive form. Then use Algorithm 1 or 2 to calculate the reliability $re'(t)$ corresponding to $e'(t)$. Then,

$$re(t) = 1 - re'(t)$$

Example 7.4.23 *Continuing with Example 7.4.22, we have*

$e(t) = f_1 \wedge (f_2 \vee \neg f_3)$

Then,

$e'(t) = \neg e(t) = \neg f_1 \vee (\neg f_2 \wedge f_3)$

Using Algorithm 1:

$$\begin{aligned} K_1 &= re(t@(\bar{1}00)) + re(t@(0\bar{1}1)) \\ &= 0.40 + (0.3)(0.80) \\ &= 0.64 \\ K_2 &= re(t@(\bar{1}\bar{1}1)) \\ &= (0.4)(0.3)(0.80) \\ &= 0.096 \\ re'(t) &= K_1 - K_2 \\ &= 0.544 \\ re(t) &= 1 - re'(t) \\ &= 1 - 0.544 = 0.456 = 45.6\% \end{aligned}$$

REFERENCES

1. Abiteboul, S., Kanellakis, P., and Grahne, G., On the representation and querying of sets of possible worlds, *Theoretical Computer Science,* **78**:159–187 (1991).

2. Aho, A., Sagiv, Y., and Ullman, J. D., Equivalences among relational expressions, *SIAM J. Computing,* **8**(2):218–246 (1979).

3. Alagar, V. S., Sadri, F., and Said J. N., Semantics of an extended relational model for managing uncertain information, *Proceedings of Fourth International Conference on Information and Knowledge Management,* pp. 234–240 (1995).

4. Apt, K. R., Introduction to logic programming, In van Leeuwen, editor, *Handbook of Theoretical Computer Science,* North Holland (1986).

5. Apt, K. R., Blair, H. A., and Walker, A., Towards a theory of declarative knowledge, In J. Minker, editor, *Foundations of Deductive Databases and Logic Programming,* Morgan-Kaufmann (1988).

6. Bancilhon, F., Maier, D., Sagiv, Y., and Ullman, J. D., Magic sets and other strange ways to implement logic programs, *Proc. ACM Symp. PODS,* pp. 1–15 (1986).

7. Barbara, D., Garcia-Molina, H., and Porter, D., The management of probabilistic data, *IEEE Transactions on Knowledge and Data Engineering,* **4**(5):487–502 (1992).

8. Chandra, A. K. and Merlin, P. M., Optimal implementation of conjunctive queries in relational databases, *Proc. 9th Annual ACM Symp. on the Theory of Computing,* pp. 77–90 (1977).

9. Fagin, R., Halpern, J., and Meggido, N., A logic for reasoning about probabilities, *Information and Computation,* (1992).

10. Garey, M. R. and Johnson, D. S., *Computers and Intractability: A Guide to the Theory of NP Completeness,* W.H. Freeman and Company, New York (1979).

11. Gelfond, M. and Lifischitz, V., The stable model semantics for logic programming, *Proc. 5th IEEE Conf. and Symposium on Logic Programming,* pp. 1071–1080, Seattle, Washington, NY (1988).

12. Güntzer, U., Kießling, W., and Thöne, H., New directions for uncertainty reasoning in deductive databases, *Proc. ACM SIGMOD Intl. Conf. on Management of Data,* pp. 178–187 (1991).

13. Iaonnidis, Y. E. and Ramakrishnan, R., Containment of conjunctive queries: Beyond relations as sets, *ACM Transactions on Database Systems,* **20**, **3**:288–324 (1995).

14. Kaufmann, A., *Introduction to the Theory of Fuzzy Sets,* Vol. 1, Academic Press, Inc., Orlando, Florida (1973).

15. Kifer, M. and Li, A., On the semantics of rule based expert systems with uncertainty. In M. Gyssens, J. Paradaens, and D. van Gucht, editors, *2nd Intl. Conf, on Database Theory,* 102–117, Springer-Verlag LNCS-326, Bruges, Belgium (1988).

16. Klir, G.J., U. St. Clair, U.H., and Yuan, B., *Fuzzy Set Theory, Foundations and Applications,* Prentice Hall, Upper Saddle River, N.J. (1997).

17. Klir, G.J. and Folger, T.A., *Fuzzy Sets, Uncertainty and Information,* Prentice Hall, Englewood Cliffs, N.J. (1988).

18. Klir, G.J. and Wierman, M.J., *Uncertainty-Based Information,*Physica-Verlag, Heidelberg, Germany (1999).

19. Klir, G.J. and Yuan, B., *Fuzzy Sets and Fuzzy Logic: Theory and Applications,* Prentice Hall, Upper Saddle River, N.J. (1995).

20. Lee, S. K., An extended relational database model for uncertain and imprecise information an evidential approach, *Proc. of the Intl. Conf. on Very Large Data Bases,* pp. 211–220 (1992).

21. Lee, S. K., Imprecise and uncertain information in databases: An evidential approach, *Proceedings of the Intl. Conf. on Data Engineering,* pp. 614–621 (1992).

22. Liu, K.C. and Sunderraman, R., Indefinite and maybe information in relational databases, *ACM Transactions on Database Systems,* **15**(1):1–39 (1990).

23. Liu, K.C. and Sunderraman, R., A generalized relational model for indefinite and maybe information, *IEEE Transactions on Knowledge and Data Engineering,* **3**(1):65–77 (1991).

24. Lloyd, J. W., *Foundations of Logic Programming,* Springer-Verlag, second edition (1987).

25. Mordeson, J. N. and Nair, P.S., *Fuzzy Mathematics,* Physica-Verlag, Heidelberg, Germany, second edition (2001).

26. Mordeson, J. N. and Nair, P.S., *Fuzzy Graphs and Fuzzy Hypergraphs,* Physica-Verlag, Heidelberg, Germany (2000).

27. Motro, A., Integrity = validity + completeness, *ACM Transactions on Database Systems,* **14**(4):480–502 (1989).

28. Ng, R. T. and Subrahmanian, V. S., A semantical framework for supporting subjective and conditional probabilities in deductive databases, *Automated Reasoning,* **10**(2):191–235 (1993).

29. Ng, R. T. and Subrahmanian, V. S., Stable semantics for probabilistic deductive databases, *Information and Computation,* **110**(1):42–83 (1994).

30. Rosenfeld, A., Fuzzy graphs, In: L. A. Zadeh, K. S. Fu and M. Shimura, editors, *Fuzzy Sets and Their Applications,* Academic Press, New York, 77–95 (1975).

31. Sadri, F., Reliability of answers to queries in relational databases, *IEEE Transactions on Knowledge and Data Engineering,* **3**(2):245–251 (1991).

32. Sadri, F., Aggregate operations in information source tracking method, *Theoretical Computer Science,* **133**:421–442 (1994).

33. Sadri, F. Information source tracking method: Efficiency issues, *IEEE Transactions on Knowledge and Data Engineering,* **7**(6):947–954 (1995).

34. Sadri, F., Integrity constraints in the information source tracking method, *IEEE Transactions on Knowledge and Data Engineering,* **7**(1):106–119 (1995).

35. Sadri, F., Modeling uncertainty in databases, *Proc. 7th IEEE Intl. Conf. on Data Eng.*, pp. 122–131 (1991).

36. Sagiv, Y., Optimizing datalog programs, In J. Minker, editor, *Foundations of Deductive Databases and Logic Programming,* pp. 659–698. Morgan-Kaufmann (1988).

37. Shmueli, O, Decidability and expressiveness aspects of logic queries, *ACM Principles of Databases Systems,* pp. 237–249 (1987).

38. Silberschatz, A., Stonebraker, M., and Ullman, J. D., Database systems: Achievements and opportunities, *Communications of the ACM,* **34**:110–120 (1991).

39. Tamura, S., Higuchi, S., and Tanaka, K., Pattern Classification Based on Fuzzy Relations, *IEEE Trans.* SMC-1, pp. 61–66 (1971).

40. Ullman, J. D., Bottom-up beats top-down for datalog, *Proc. ACM Symp. PODS,* pp. 140–149 (1989).

41. Ullman, J. D., *Principles of Database and Knowledge Base Systems,* volumes I & II, Computer Science Press, Maryland (1989).

42. van Emden, M. H., Quantitative deduction and its fixpoint theory, *Journal of Logic Programming,* **4**(1):37–53 (1986).

43. van Emden, M. H. and Kowalski, R. A., The semantics of predicate logic as a programming language, *JACM,* **23**(4):733–742 (1976).

44. van Gelder, A., Ross, K., and Schlipf, John S., Unfounded sets and well founded semantics of general logic programs, *Proc. 7th ACM Symposium on Principles of Database Systems,* pp. 221–230, Austin, Texas. New York (1988).

45. Yeh, R. T. and Bang, S.Y., Fuzzy graphs, fuzzy relations, and their applications to cluster analysis. In: L. A. Zadeh, K. S. Fu and M. Shimura, editors, *Fuzzy Sets and Their Applications,* Academic Press, New York, pp. 125–149 (1975).

46. Zadeh, L. A., Fuzzy sets as a basis for a theory of possibility, *Fuzzy Sets and Systems,* **1**:3–28 (1978).

47. Zadeh, L.A., Similarity relations and fuzzy orderings, *Information Sci.,* **3**:177–200 (1971).

48. Zimmermann, H.J., *Fuzzy Set Theory and Its Applications,* second edition, Kluwer Academic Publishers, Boston (1991).

LIST OF FIGURES

LIST OF TABLES

INDEX